Assessment of Waters for Estrogenic Activity

The mission of the Awwa Research Foundation (AwwaRF) is to advance the science of water to improve the quality of life. Funded primarily through annual subscription payments from over 1,000 utilities, consulting firms, and manufacturers in North America and abroad, AwwaRF sponsors research on all aspects of drinking water, including supply and resources, treatment, monitoring and analysis, distribution, management, and health effects.

From its headquarters in Denver, Colorado, the AwwaRF staff directs and supports the efforts of over 500 volunteers, who are the heart of the research program. These volunteers, serving on various boards and committees, use their expertise to select and monitor research studies to benefit the entire drinking water community.

Research findings are disseminated through a number of technology transfer activities, including research reports, conferences, videotape summaries, and periodicals.

Assessment of Waters for Estrogenic Activity

Prepared by:
Jocelyn D.C. Hemming, Miel A.E. Barman,
and **Jon H. Standridge**
Wisconsin State Laboratory of Hygiene
2601 Agriculture Drive
Madison, WI 53707-7996

Sponsored by:
Awwa Research Foundation
6666 West Quincy Avenue
Denver, CO 80235-3098

Published by the

DISCLAIMER

This study was funded by the Awwa Research Foundation (AwwaRF). AwwaRF assumes no responsibility for the content of the research study reported in this publication or for the opinions or statements of fact expressed in the report. The mention of trade names for commercial products does not represent or imply the approval or endorsement of AwwaRF. This report is presented solely for informational purposes.

Published by IWA Publishing, Alliance House, 12 Caxton Street, London SW1H 0QS, UK
Telephone: +44 (0) 20 7654 5500; Fax: +44 (0) 20 7654 5555; Email: publications@iwap.co.uk;
Web: www.iwapublishing.com

AwwaRF report number 90940F. Originally published by AwwaRF for its subscribers in 2003.
IWA Publishing version published 2004.

Copyright © 2003
by
Awwa Research Foundation

ISBN 1 84339 852 4

CONTENTS

LIST OF TABLES vii

LIST OF FIGURES ix

FOREWORD xi

ACKNOWLEDGMENTS xiii

EXECUTIVE SUMMARY xv

CHAPTER 1: INTRODUCTION 1
    Background Information 1
    Objectives 3

CHAPTER 2: METHODS AND MATERIALS 5
    In Vitro E-Screen 5
        Sample Collection and Extraction 5
        MCF-7 Cell Culturing 5
        Proliferation Assay (E-Screen) 6
    In Vivo Fish Studies 7
        In-lab Fish Studies 7
        Caged Fish Studies 9
    Statistical Analyses 9
        Fish Studies 9

CHAPTER 3: RESULTS AND DISCUSSION 11
    E-Screen 11
        Standard Curve 11
        Results of Blanks and Spikes 12
        Results from Drinking Water Facilities 13
        Results from Wastewater Treatment Plant Effluents 18
        Evaluation of E-Screen as Tool 19
    Fish Studies 20
        Lab Dose Response Study 20
        Exposure of Fish to Facility Waters 20
    Comparison of E-Screen and Fish Responses 21
    Conclusions 22

CHAPTER 4: SUMMARY AND CONCLUSIONS 23
    Project Description 23
    Purpose 23
    Approach 23
    Significant Results 24

Conclusions	24
CHAPTER 5: RECOMMENDATIONS FOR THE WATER INDUSTRY	25
APPENDIX A: SAMPLE COLLECTION FORM	27
APPENDIX B: E-SCREEN EC_{50} RUN CHART	29
APPENDIX C: E-SCREEN RESULTS FOR SOURCE AND FINISHED WATERS	31
APPENDIX D: E-SCREEN RESULTS FOR WWTP EFFLUENTS	49
APPENDIX E: MALE FATHEAD MINNOW EXPOSURES	55
REFERENCES	61
ABBREVIATIONS	65

TABLES

3.1	E-Screen results from 2.7 ng/L 17β-estradiol-spiked samples	12
3.2	E-Screen results from duplicate samples	13
3.3	E-Screen activity in estradiol equivalents in samples of source and finished waters	16
3.4	E-Screen activity in estradiol equivalents for various wastewater effluents prior to and after water reclamation processes	19
C.1	Facility information and E-screen activity of all source and finished waters tested	33
D.1	Facility information and E-screen activity of all wastewater treatment plant effluents tested	51
E.1	Responses of male fathead minnows to different waters	56

FIGURES

3.1	A typical E-Screen standard curve	11
3.2	Number of finished waters tested with the E-Screen results indicated	14
3.3	Number of source waters tested with the E-Screen results indicated	14
3.4	Number of wastewater treatment plant effluents tested with the E-Screen results indicated	18
3.5	Plasma vitellogenin concentrations for adult male fathead minnows exposed to varying concentrations of estradiol for twenty-one days	20
B.1	The run chart of EC_{50} from the E-Screen assays	30

FOREWORD

The Awwa Research Foundation is a nonprofit corporation that is dedicated to the implementation of a research effort to help utilities respond to regulatory requirements and traditional high-priority concerns of the industry. The research agenda is developed through a process of consultation with subscribers and drinking water professionals. Under the umbrella of a Strategic Research Plan, the Research Advisory Council prioritizes the suggested projects based upon current and future needs, applicability, and past work; the recommendations are forwarded to the Board of Trustees for final selection. The foundation also sponsors research projects through the unsolicited proposal process; the Collaborative Research, Research Applications, and Tailored Collaboration programs; and various joint research efforts with organizations such as the U.S. Environmental Protection Agency, the U.S. Bureau of Reclamation, and the Association of California Water Agencies.

This publication is a result of one of these sponsored studies, and it is hoped that its findings will be applied in communities throughout the world. The following report serves not only as a means of communicating the results of the water industry's centralized research program but also as a tool to enlist the further support of the nonmember utilities and individuals.

Projects are managed closely from their inception to the final report by the foundation's staff and large cadre of volunteers who willingly contribute their time and expertise. The foundation serves a planning and management function and awards contracts to other institutions such as water utilities, universities, and engineering firms. The funding for this research effort comes primarily from the Subscription Program, through which water utilities subscribe to the research program and make an annual payment proportionate to the volume of water they deliver and consultants and manufacturers subscribe based on their annual billings. The program offers a cost-effective and fair method for funding research in the public interest.

A broad spectrum of water supply issues is addressed by the foundation's research agenda: resources, treatment and operations, distribution and storage, water quality and analysis, toxicology, economics, and management. The ultimate purpose of the coordinated effort is to assist water suppliers to provide the highest possible quality of water economically and reliably. The true benefits are realized when the results are implemented at the utility level. The foundations trustees are pleased to offer this publication as a contribution toward that end.

Edmund G. Archuleta, P.E.
Chair, Board of Trustees
Awwa Research Foundation

James F. Manwaring, P.E.
Executive Director
Awwa Research Foundation

ACKNOWLEDGMENTS

The authors of this report would like to express our appreciation to the many water utilities that provided samples for this project.

The research team is also grateful to the following people for their assistance in completing this project: members of the WSLH biomonitoring team including Steve Geis, Eric Herro, Dawn Karner, Amy Mager, Karen Schappe, Gwen Schaumberger, and Xiaoxia Yu, and Bette Weishhaar for assistance with cell culturing.

We are also indebted to Jim Amrhein and Jim Killian from the Wisconsin Department of Natural Resources for their invaluable help in conducting the fish exposure studies.

In addition, the authors are grateful to Drs. Ana Soto and Carlos Sonnenschein and the members of their lab (Nancy Prechtl, Janine Calabro and Cheryl Michaelson) for the donation of the MCF-7 cells, and for the training and advice received from their lab.

The authors also would like to thank Nancy Denslow and Kevin Kroll of the University of Florida in Gainesville for running the vitellogenin ELISA samples.

Finally, we wish to thank the AwwaRF project managers Jarmila Popovicova, Kim Linton and Misha Hasan, Project Advisory Committee members, Djanette Khiari, Tim Kubiak, Yves Levi and Shane Snyder for their guidance and advice during this project.

EXECUTIVE SUMMARY

Recently, scientific evidence has alerted public health officials, environmental scientists and the drinking water industry that diverse classes of chemicals found in the environment can disrupt the endocrine systems of fish, wildlife, and possibly humans, resulting in measurable, adverse health effects. For example, the decline of the alligator population in Lake Apopka, Florida, was linked to changes in alligator hormone levels following pesticide spills. Hermaphroditic fish have been discovered downstream of sewage-treatment works in the United Kingdom. And male fish collected from a side channel of the Mississippi River and in the Las Vegas Wash have exhibited elevated vitellogenin levels, indicating the presence of estrogenic compounds in waterways used by communities as sources for their public water supplies. A diverse array of man-made and natural chemicals are characterized as estrogenic, i.e. compounds that produce the same effects as naturally occurring estrogens. Compounds have been identified as estrogenic using tests that indicate the compound binds to the estrogen receptor and elicits a response. Many of these estrogenic compounds are found in aquatic systems, including alkylphenol polyethoxylates (nonionic surfactants used in the production of detergents, plastics and agricultural chemicals), phthalates (used in the production of various plastics including PVC), natural estrogens (i.e., 17β-estradiol and estrone) and the synthetic estrogen 17α-ethynylestradiol (an active component of the oral contraceptive pill), β-sitosterol, (a natural plant compound released during wood processing), some organochlorine pesticides and industrial chemicals such as dioxin and PCBs. Consequently, the USEPA has deemed the study of endocrine disrupters a high priority. It is highly likely that regulations concerning the discharge of these compounds will be considered. It is imperative that the drinking water community, have as much information as possible about the occurrence and significance of these compounds so as to effectively participate in the regulatory rule making process.

PURPOSE

The goal of this research was to determine the occurrence, prevalence and significance of estrogenic activity in source waters, finished drinking waters and industrial and municipal wastewater effluents impacting source waters. This goal was accomplished by fulfilling the following objectives.

1. Validate and optimize the E-Screen assay for use with water samples.
2. To document the prevalence of estrogenic activity in waters used as sources for drinking waters, the finished drinking waters, and WWTP effluents.
3. Conduct caged fish studies, in lab exposure studies and the E-Screen assay on the same samples.
4. Evaluate the test results: Assess the reliability of a rapid screen assay compared to in vivo assays. Postulate the possible public health significance of the findings and suggest further research needs.

APPROACH

In this research, two biological assays were used to indicate whether waters contained compounds with estrogenic activity. One was the E-Screen, a cell culture based assay in which a breast cancer cell line proliferates in response to estrogenic compounds. In the other, male fish exposed to estrogenic compounds induced vitellogenin, a precursor to egg yolk protein, normally only found in females. First, the E-screen was optimized for use with water samples, and then over 70 drinking water facilities were screened. Source water, finished water and some effluents from WWTPs were tested. Additionally, caged male fathead minnows were exposed in situ to source waters and if appropriate, effluents from an impacting contributor. Male fathead minnow exposures were also conducted in the lab to source water, finished drinking water and WWTP effluents. The fish studies were conducted at six facilities.

RESULTS

Using the E-Screen, we tested 90 finished water samples from 72 facilities. The vast majority of the finished water samples (84%) did not contain estrogenic activity above the level of detection for this assay (0.029 ng/L estradiol equivalents). Of the remaining 14, 13 were below 0.27 ng/L. In contrast to the finished drinking waters, the majority (61%) of source waters did have measurable estrogenic activity. Of the 105 surface or ground water samples tested, 64 were above the detection limit of the assay. Again, the activities were fairly low, with only 10 waters having activities above 0.27 ng/L. Not surprisingly, surface waters exhibited higher levels of estrogenic activity than did ground waters. What was surprising was that 42% of the ground waters tested had estrogenic activity. Twenty-six wastewater treatment plant effluents were tested from 16 facilities. Estrogenic activity ranged from no activity (in seven of the effluents) to a sample with greater than 500 ng/L of activity.

Fish exposure studies were completed at six facilities. Vitellogenin levels were significantly increased in male fathead minnows in only two of the waters tested, both wastewater treatment plant effluents. No significant responses were found in fish exposed to the source or drinking water samples. Comparison between the E-Screen and vitellogenin responses were limited by the low number of positive vitellogenin responses, although based on the E-Screen results, fish would have been expected to respond after exposure to one additional effluent.

CONCLUSIONS

The present research indicates that the E-Screen is a very effective tool for assessing low levels of in vitro estrogen activity in source and finished waters. The sensitivity and relatively short duration of this assay makes it a more useful assay than fish exposures, which are impractical as a lab screening assay due to the long duration and large volumes of water needed. It was striking that such a high percentage of source waters exhibited measurable in vitro estrogenic activity, and reassuring that drinking water treatments reduced activity, usually to levels below detection. This study and others indicate that wastewater treatment plant effluents are one likely source for these compounds. Neither the E-Screen or the induction of vitellogenin can indicate what specific compounds are responsible for this activity. However, natural and synthetic estrogens have been reported to be primarily responsible for estrogenic activity in

WWTP effluents. The levels seen in a few of the highly active WWTP effluents in this study may be high enough to affect the health of aquatic organisms. Whether the low levels, such as those seen in a drinking water samples are high enough above human background levels to have an effect is not known. Research is needed to determine if there is a health risk from chronic low levels of estrogenic compounds. Additionally, research is needed to identify specific compounds responsible for the activity and to determine which processes can remove these compounds during treatment at WWTP and drinking water facilities.

CHAPTER 1
INTRODUCTION

BACKGROUND INFORMATION

Emerging evidence indicates that diverse classes of chemicals in the environment disrupt the endocrine systems of fish and wildlife. For example, the decline of the alligator population in Lake Apopka, Florida, was linked to changes in alligator hormone levels after a pesticide mix (containing mainly dicofol and some DDT) was spilled (Guillette et al.1994; Guillette et al.1996). Female alligators from the contaminated lake had twice the normal levels of estrogen and male alligators had depressed testosterone levels, similar to levels found in females at the control site. These hormonal changes were accompanied by morphologic abnormalities in their gonads and phalli. Researchers in the United Kingdom discovered hermaphroditic fish downstream of a sewage-treatment works (Sumpter and Jobling 1995), which lead to the concern that endocrine disruption is a wide-spread problem in aquatic ecosystems. Extensive work in Europe documented that male fish exposed to effluents from sewage treatment works (Sumpter and Jobling 1995, Harries et al. 1996, Harries et al. 1997, Knudsen et al. 1997, Lye et al. 1998), pulp and paper mills (Soimasuo et al. 1998) and oil refinery treatment works (Knudsen et al. 1997) contained high levels of a protein generally synthesized only by females called vitellogenin. Additionally, the incidence of intersex (gonads with simultaneous male and female characteristics) in wild collected roach was especially high in wastewater treatment plant (WWTP) effluent impacted rivers (Jobling et al. 1998). In the United States, male fish from a side channel of the Mississippi River receiving effluent from the St. Paul Metropolitan Sewage Treatment Works (Folmar et al. 1996, 2001) and in the Las Vegas Wash in Utah (Roefer et al. 2000), also exhibited elevated vitellogenin levels.

Because of results such as these and the controversial worries that endocrine disrupters may be affecting human health (e.g., declining sperm counts and increasing rates of cryptorchidism, hypospadias, endometriosis, and breast, prostate, and testicular cancers [Sharpe and Skakkebaek 1993, Jensen et al. 1995, Colburn et al. 1996, but see e.g. Safe 2000, Handelsman 2001]), public and environmental health scientists and regulators are looking closely at this new research. In fact, EPA has deemed the study of endocrine disrupters a high priority (US EPA, 1997). It is very likely that regulations concerning the occurrence of these compounds will be promulgated. It is imperative that the drinking water community have as much information as possible about the occurrence and significance of these compounds so as to effectively participate in any regulatory rule making processes.

Much concern has focused on compounds with estrogenic activity – that is, compounds that produce the same effects as naturally occurring estrogens. Estrogens are primarily involved in the regulation of sexual development and reproduction in female vertebrates, although estrogens also play a role in some metabolic processes including growth. Compounds are considered estrogenic if they bind to estrogen receptors in cells and interact with estrogen responsive elements in the DNA, leading to expression of appropriate genes.

Compounds with estrogenic activity can be detected with several different types of in vivo and in vitro assays (EDSTAC 1998). With in vivo tests, live organisms are first exposed to compounds and then are assessed for responses typical of exposure to estrogen. In mammals, this usually involves analysis of the uterus. In fish, the measure of vitellogenin in male plasma is a sensitive endpoint. Estrogens stimulate the liver to produce vitellogenin, a precursor to egg yolk, and vitellogenin is normally found in appreciable levels in females only. However, vitellogenin synthesis is induced in males in response to estrogens or compounds with estrogenic activity and is therefore a widely-used biomarker (e.g., Sumpter and Jobling, 1995, Panter et al.1998, Jones et al. 2000). The growth and maintenance of gonads is also under hormonal control. Thus, testes weight relative to body weight (i.e., gonadosomatic index [GSI]) is indicative of exposure to estrogenic compounds, although it is not as sensitive as vitellogenin induction (Jobling et al.1996, Harries et al.1997). With in vitro tests, cells or extracts of cells are exposed to compounds and then are assessed for evidence of receptor binding. These include tests that can determine the affinity of a compound for the estrogen receptor (competitive ligand binding), or endpoints that indicate transcription of estrogen responsive genes. One commonly used in vitro assay is the E-Screen (Soto et al.1995), which was developed as a screening tool to detect potentially estrogenic compounds. Estrogens cause MCF-7 cells, a human breast cancer cell line, to proliferate. This proliferation is an estrogen receptor mediated process and therefore proliferation indicates exposure to estrogenic compounds.

Based on the assays described above, a diverse array of man-made and natural chemicals are characterized as estrogenic, many of which are found in aquatic systems. Alkylphenol polyethoxylates are nonionic surfactants used in the production of detergents, plastics and agricultural chemicals. Microbial degradation during the sewage treatment process results in smaller, estrogenic metabolites, such as the alkylphenols nonylphenol and octylphenol, which are found in appreciable levels in the aquatic ecosystem, at levels known to have estrogenic effects (Jobling and Sumpter 1993, Soto et al. 1995, Jobling et al. 1996, Tyler et al. 1998 and references therein). Phthalates are plasticizers used to give plastics flexibility. Phthalate esters, used in the production of various plastics (including PVC), are among the most ubiquitous industrial chemicals in the environment. *In vitro* tests have determined that phthalates are weakly estrogenic (Jobling et al. 1995, Soto et al. 1995), however, not enough *in vivo* work has been done to know if concentrations are high enough to have ecological consequences (Tyler et al. 1998). Bisphenol A, a monomer used in the manufacturing polycarbonate, is also estrogenic (Routledege and Sumpter 1996). Natural estrogens (i.e., 17β-estradiol and estrone) and the synthetic estrogen 17α-ethynylestradiol (an active component of the oral contraceptive pill) were identified as the compounds having estrogenic activity in effluents from sewage treatment plants (Desbrow et al. 1998, Matsui et al. 2000, Snyder et al. 2001). β-sitosterol, a natural plant compound released during wood processing (and found in bleached pulp mill effluent) has estrogenic effects in fish (Mellanen et al.1996). Finally, some organochlorine pesticides and industrial chemicals such as dioxin and PCBs exhibit estrogenic activity (Soto et al.1995; Tyler et al.1998).

The Safe Drinking Water Act (PL 104-170) and the Food Quality Protection Act (PL 104-182) were amended in 1996 to include a mandate for the U.S. EPA to develop testing programs to identify endocrine disrupting compounds. Because of this, thousands of individual compounds will be screened for potential endocrine disrupting effects (EDSTAC 1998, Zacharewski 1997). In reality, however, aquatic organisms will be exposed to the complex mixtures of compounds in effluents and the waters they impact. It is still unclear whether endocrine disrupters act in an additive, synergistic or antagonistic manner (Crisp et al.1998), and thus assessment of the estrogenic activity is important. To assess potential risk, it is important to look at diverse samples of waters so that the scope of the problem is documented. Work has been done in Europe, but research in the United States concerning the distribution of waters with estrogenic activity has only recently been undertaken (Rudel et al. 1998, Snyder et al. 1999, Kolpin et al. 2002).

Thus, the purpose of this research was to determine the prevalence of estrogenic activity in a variety of water samples using an in vitro and in vivo test method. The in vitro test assay, the E-screen, allowed us to sample a wide variety of water samples in a cost effective manner. In vitro tests, however, provide little information regarding the ecological consequences of estrogenic activity. In vivo tests are more realistic (e.g., they rely on normal routes of exposure and allow for metabolic activation), but the time and expense of conducting in vivo assay limits testing to high priority samples only. The in vivo studies involved exposing male fathead minnows to samples and measuring changes in vitellogenin levels and gonadosomatic index, both characteristics of exposure to estrogenic compounds. The in vivo study were conducted in situ, with fish caged in source water, simultaneous with fish exposed to water samples collected and shipped to the laboratory. Results of the in vivo and in vitro results were compared in order to assess the degree to which screenings with the E-screen tool correlated with observed estrogenic effects in the fish experiments.

OBJECTIVES

The goal of this research was to determine the occurrence, prevalence and significance of estrogenic activity in source waters, finished drinking waters and industrial and municipal wastewater effluents impacting source waters. This goal was accomplished by fulfilling the following objectives.

1. Validate and optimize the E-Screen assay for use with water samples.
2. To document the prevalence of estrogenic activity in waters used as sources for drinking waters, the finished drinking waters, and WWTP effluents.
3. Conduct caged fish studies, in lab exposure studies and the E-Screen assay on the same samples.
4. Evaluate the test results: Assess the reliability of a rapid screen assay compared to in vivo assays. Postulate the possible public health significance of the findings and suggest further research needs.

CHAPTER 2
METHODS AND MATERIALS

IN VITRO E-SCREEN

Sample Collection and Extraction

Interested utilities were recruited to participate in this study through Awwa publications. The large number of volunteers demonstrates the water industries willingness to stay ahead of emerging environmental contamination issues. A copy of the lab slip used to gather information about the samples is included in Appendix A. Samples of source waters, finished waters and WWTP effluents were collected by facility operators in 1-L amber bottles and shipped on ice or with pre-frozen cold packs to the laboratory via an overnight courier. If samples were not immediately extracted, they were placed at 4°C until extraction, usually within 48 hours of receipt. Surface water and WWTP effluent samples were pre-filtered through a muffled (450°F) glass fiber filter. The water samples were then extracted using two 3M Empore™ (St. Paul, Minn.) 47 mm extraction disks: a C18 and a SDB-XC disk. The method for the C18 disk was based on EPA Method 525.1. A C18 disk was washed with 10 mL of a 1:1 dichloromethane (DCM): ethyl acetate mix. The disk was then conditioned with 5 mL methanol, followed by a 20 mL type I water rinse. After conditioning, a 1-L water sample was pulled through the disk and eluted first with ethyl acetate, then a 1:1 mixture of ethyl acetate and DCM, followed by DCM alone. The method for the SDB-XC disk was based on EPA quick turnaround method for phenols. The SDB disk was swelled and pre-washed with acetone followed by isopropyl alcohol. The disk was then washed with 10 mL of DCM. Finally, the disk was conditioned with 5 mL of methanol followed by a 20 mL type I water rinse. The 1-L sample (previously run through the C18 disk) was modified by adding 50g NaCl and lowering the pH to 2 with DCM cleaned HCl. The sample was pulled through the disk and then the disk was eluted with 3 5-mL rinses of DCM. The extracts from each disk were combined and blown to near dryness with nitrogen. The extracts were transferred with several ethanol rinses to a calibrated 2-mL amber vial, dried to near dryness and brought back up to 1 mL in ethanol. Samples were stored in a 4°C cooler until use in the assay.

MCF-7 Cell Culturing

A breast cancer cell line (MCF-7 cells) was continuously cultured for use in the E-Screen. The cells were obtained from Drs. Sonnenschein and Soto at Tufts University (Boston, Mass.). The MCF-7 cells were cultured in Dulbecco's modified eagle's medium (ICN Biomedicals, Aurora, Ohio) with 5% fetal bovine serum (FBS) (Hyclone Laboratories, Logan, Utah) at 37°C and 6.5% CO_2 in 75 cm^2 tissue culture flasks. Media was changed every 2-3 days and the cells are subcultured every seven days. Stocks of MCF-7 cells were stored in liquid nitrogen and new cells were thawed to replace cells that had undergone approximately 20 passages.

Proliferation Assay (E-Screen)

For the E-Screen, seven days following sub-culturing, cells from a 75-cm^2 flask were trypsinized and counted on an EPICS XL flow cytometer (Coulter Corporation, Miami, Fla.) Cell counts were achieved using FlowCount beads (Coulter Corporation, Miami, Fla.) an FDA-approved microbead standard originally developed for enumeration of blood cells in human patients. A flow cytometer is a laser-based instrument that can rapidly analyze thousands of individual cells or nuclei in a suspension. The cells are suspended in a fine stream so that each individual cell passes through the flow cytometer's laser beam. The instrument measures the amount of light scattered in the forward and in the 90° angle direction from each cell. These measurements correlate to size of each cell and are done at the rate of thousands of cells per second. These signals are collected plotted on user-defined histograms. Cells with the specific light scatter characteristics of the target cells or nuclei appear as data points in the defined area of the histogram and are electronically enumerated. Once the concentration of the cells was determined, cells were then seeded in 24-well plates to achieve 20,000-30,000 cells per well, in 1-mL of culturing media. Twenty-four hours after seeding, the media was removed and experimental media was added.

The experimental media was Dulbecco's Modification of Eagles Medium without phenol red (Irvine Scientific, Irvine Calif.). The 5% FBS used in this media was stripped of steroids with a charcoal dextran (CD) stripping procedure described in Payne et al. (2000). Briefly, FBS was incubated with activated charcoal and dextran followed by centrifugation and filtration (0.2μm) to remove the charcoal. This experimental media is referred to as CD-media.

For plates containing the standards, the MCF-7 cells were exposed to 15 concentrations of 17β-estradiol (Sigma Chemical Co. St. Louis Mo.) in CD-media ranging from 0.14 to 2724 ng/L. All treatments were done in quadruplicate, with three plates needed to obtain a complete standard curve. Four control wells were included on each plate. For plates on which samples were run, 0.1 mL of the water extract (in ethanol) was added to 9.9 mLs of CD-media. A 50% dilution series was then made for a total of 5 concentrations. Each concentration was applied to four wells at a volume of 0.5 mLs. In one of those wells, estradiol was added to obtain a concentration of 27.2 ng/L in the well. This positive control was used to determine whether anything in the sample was preventing the growth of the cells. Again, each plate contained 4 control wells.

After five days of incubation, cell proliferation was measured. For the estradiol standard curve, cell proliferation was measured by both flow cytometry and the sulphorhodamine B (SRB) (Sigma Chemical Co. St. Louis Mo.) protein assay. The flow cytometry counts were done to enable correlations between the cell number and optical density (OD) results from the SRB assay. The cells were lysed (Soto et al. 1995) and nuclei were counted with the flow cytometer. The SRB assay determines total cell numbers by measuring total protein content and is much less expensive. This procedure was used following the five day experimental exposure, at which time the media was removed from the cells and 10% trichloroacetic acid (TCA) solution was added to each well to fix the cells. After 20-30 minutes, the TCA was removed, the wells were allowed to dry and the SRB solution (0.4% SRB dye in 1% acetic acid/99% distilled water) was added to each well to stain the cells. The residual SRB dye was removed by rinsing with 1% acetic acid after 20-30 minutes. The remaining dye was redissolved using 10mM Tris solution (pH 10.5) and read at a wavelength of 515 nm. Proliferation was measured by the SRB assay only for the plates containing water sample tests.

To determine the estradiol equivalents (Eeq) of samples, the standard curve was fit with a 4-parameter logistic equation. This is done with the Softmax PRO v. 2.6 software package, the software analysis package for the Molecular Devices microplate reader (Sunnyvale Calif.). The estradiol equivalents for unknown samples were calculated from this equation. The estradiol equivalent was corrected for the dilution of the sample for use in the E-screen (at least 100-fold), and the concentration of the water sample during the extraction procedure (1000-fold). Dilutions resulting with values closest to the concentration which caused 50% of the maximum proliferation (EC_{50}) for the standard curve were reported and standard deviations were calculated from the three replicate wells in an individual assay. The limit of detection (LOD) of this assay was determined by running seven extracts from 1-L Milli-Q water blanks and multiplying the standard deviation of the results by 3.29 (APHA et al. 1992). This value was added to the lower asymptote of the standard curve and was determined to be 0.029 ng/L estradiol. The limit of quantitation was calculated to be ten times the standard deviation above the lower asymptote and was determined to be 0.079 ng/L. The method detection limit for this assay was not determined. The EC_{50} for each standard curve was recorded to ensure the cells were responding consistently (Figure B.1).

IN VIVO FISH STUDIES

Fathead minnows used in the experiments were from the culture maintained at the Wisconsin State Laboratory of Hygiene. The culture is maintained daily according to the guidelines established by USEPA (USEPA, 1987) with slight modifications including a recirculating water supply system. The minnows' age, health, diet and environment are monitored on a continuous basis. Males used in the study were put in male-only tanks as soon as the sex could be determined.

In-lab Fish Studies

Adult male dose response exposure to estradiol

To ensure that the culture of male fathead minnows were responsive to estrogen, the fish were exposed to 17β-estradiol at 5 concentrations (0, 10, 50, 100, and 500 ng/L) for 21 days. The exposures were conducted in a 25°C walk-in incubator. Separate stock solutions of estradiol in ethanol were made for each concentration so that the volume of ethanol would be the same in each tank (0.5 μl ethanol/L water). The 0 concentrations include both dechlorinated water (lab control), and dechlorinated water + ethanol. Each of the concentrations and controls were set in triplicate 20 L tanks. Each tank contains 5 to 6 five month-old male fathead minnows and 16 L of the appropriate test water. The tanks were renewed daily with 8 L of the appropriate concentrations and controls. Dissolved oxygen (DO) and pH were monitored daily. After 21 days, each fish was anaesthetized with 3-Aminobenzoic acid ethyl ester (Sigma Chemical Co. St. Louis Mo.) at a rate of 0.5 g/L, the tail was cut off and blood was collected from the caudal vein using a heparinized capillary tube. The hematocrit tubes were centrifuged for 15 minutes at 4°C. Plasma from two fish in the same replicate tank was pooled to obtain three sub samples from each tank. The plasma was flash frozen in liquid nitrogen and stored at -80° C until shipment. The plasma was shipped via an overnight courier, on dry ice to the University of Florida for vitellogenin analysis. Gonads from each fish were removed, and were dried along with the remainder of the fish in a 110°C oven. The fish and gonad weights were used to determine the gonadosomatic index for each fish expressed as the percentage weight of the testes relative to the whole fish.

Validation of estrogen dosing

The exposure water was tested for 17β-estradiol using an estradiol enzyme linked immunosorbent assay (ELISA) kit from Cayman Chemical (Ann Arbor, Mich). The tests were conducted as described by the manufacturer. We tested the water twice during the study, one week apart. For the initial water samples, the water was collected soon after dosing, and each water was assayed in duplicate. In the first test, we assayed the final water from the tanks approximately 24 hours after the renewal. The water from each replicate tank was tested in duplicate. The limit of detection of this assay is about 8 ng/L. For the first sample of initial water, the concentrations were below the limit of detection for the 0 and 10 ng/L treatments, and 46 ± 10, 108 ± 20, and 502 ± 169 (mean $\pm$ standard deviation) for 50, 100 and 500 ng/L treatments, respectively. The tanks 24 hours after renewal were below the limit of detection except for the 500 ng/L tank, which averaged 61 ± 34 ng/L. The change in levels of estrogen over time could be due to uptake by the fish, adhesion to the tank, degradation, or problems with the assay using "fish" water. For the second sample of initial water, the concentrations were below the limit of detection for the 0 ng/L tanks, 19 ± 8, 62 ± 5, 124 ± 22 and 569 ± 110, for the 10, 50, 100 and 500 ng/L treatments, respectively.

Vitellogenin analysis

The measurement of plasma vitellogenin by direct ELISA at University of Florida was done as follows: The egg yolk precursor, vitellogenin, was quantified by direct ELISA using a monoclonal antibody. This antibody was made against carp plasma Vtg purified by anion exchange chromatography using the BIOCAD Sprint Perfusion system and POROS 20HQ media (Perseptive Biosystems, Inc.) at pH 9.0. Specificity was verified by Western analysis and ELISA against male plasma and purified vtg. Fathead minnow vtg is used to make the standard curve.

The direct ELISA was performed as follows: Plasma was diluted to several concentrations (eg. 1:100, 1:10k, 1:100k) with phosphate buffered saline containing azide and protease inhibitor- Aprotinin (PBSZ-AP), then loaded onto a polystyrene ELISA plate (NUNC brand) in triplicate. Vitellogenin standards ranging from 0.005 µg/mL to 1.0 µg/mL and positive interassay controls were also loaded onto the ELISA plate. Male plasma had been added to the standards at the same concentration as the unknowns to account for the protein effect. After incubation overnight (at 4°C), the plate was washed four times with tris-buffered saline with tween and azide (TBSTZ), and blocked for 2 hours with 1%BSA in TBSTZ with Aprotinin at room temperature. The plate was rewashed with the TBSTZ buffer and the anti-fathead vitellogenin monoclonal antibody added to each well. The quantity of the monoclonal antibody added has been previously optimized for each dilution factor used. After an overnight incubation at 4°C, the plate was washed as described previously, the secondary antibody (biotinylated goat anti-mouse) added, and incubated for 2 hours at room temperature. After, rewashing, strep-avidin conjugated to alkaline phosphatase was added to each well and incubated for 2 hours at room temperature. After a final wash with TBSTZ, the enzyme substrate (p-nitro-phenyl phosphate) was added and the color of each well measured at 405nm by an ELISA plate reader (SpectraMax) after 10-30 minutes of development. The data was interpreted using Softmax Pro program. The coefficient of variation and correlation for this assay were 10% or less and 95% of greater, respectively. The limit of detection for the vitellogenin assay is 0.5 µg/mL.

Exposure of fish to facility waters

For in-lab fish exposures to water from drinking water facilities, the utilities collected water samples on Monday, Wednesday and Fridays. When possible, upstream wastewater treatment plants also collected effluent samples. The samples were shipped to the lab *via* a courier service to arrive the following day. Samples were stored in the 4° C cooler until their use. Samples were put in a carboy, aerated and warmed to room temperature by using a aquarium heater. Because the finished drinking water samples were chlorinated, sodium thiosulfate was added (25 mL/L of a 100g/L solution). A dechlorinated tap water treatment containing sodium thiosulfate was added as an additional control. Because the fish are very susceptible to ammonia toxicity when the pH is above 8.5, HCl was added as necessary to the water to keep the pH below 8. Because the samples were shipped in polyethylene cubitainers, three control tanks with dechlorinated tap water stored in the cubitainers overnight before use were used as an additional contol. During the first summer's exposures, eighteen tanks were used; three for finished drinking water, three for the source water, three for the effluent (when appropriate), three for the cubitainer control, three for lab control, two for sodium thiosulfate control and one positive control with 100 ng/L estradiol (during the first exposure, all the fish died in a 500 ng/L estradiol tank, so 100 ng/L was used as the postive control for the remainder of the study). During the second summer's expsosures, the lab control tanks were omitted, and only the lab water stored in cubitainers were used. An additional positive control tank was included for a total of two. Renewal and processing of fish is as described for the adult male estradiol exposure study above.

Caged Fish Studies

For the in situ studies, 3 cages containing 10 fish each were placed as close to midway in the water column as feasible, using a buoy and an anchor. The cages were placed near intakes for the drinking water facilities, and the effluent outfall at the wastewater treatment facilities. The fish were retrieved 3 weeks later, and brought back to the lab for processing as described above.

STATISTICAL ANALYSES

Fish Studies.

The mean vitellogenin and GSI response from each tank or cage were analyzed with analysis of variance followed by the Student Newman Kuhls test for means separation ($\alpha = 0.05$) (PROC GLM, SAS version for Windows v. 8, Cary, N.C.). Vitellogenin results were log transformed prior to analysis.

CHAPTER 3
RESULTS AND DISCUSSION

E-SCREEN

Standard Curve

A standard curve was completed for each batch of samples analyzed (Figure 3.1). The proliferative response of the cells was sigmoidal; growth of the cells was generally not significantly higher than the controls until the estradiol concentration was greater than 0.29 ng/L and the highest proliferative response occurs at about 27.2 ng/L. Concentrations above that will not cause any increase in cell number. The EC_{50} was calculated for each run of the standard curve to ensure that responses of the cells were consistent (Figure B.1).

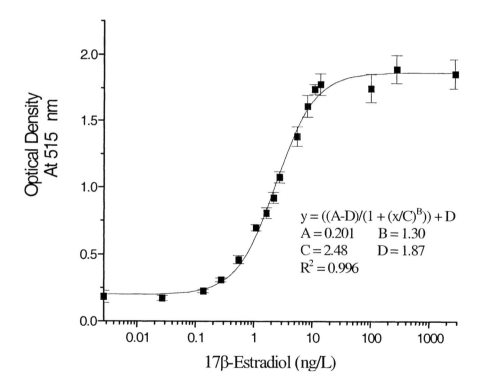

Figure 3.1 A typical E-Screen standard curve. Proliferation of MCF-7 cells after a 5-day exposure to varying concentrations of 17β-estradiol. Standard deviations represent responses from three to four wells in a single assay. The four parameter logistics equation is included on the figure. The EC_{50} is the C term from the equation.

Three out of the four standard curves in which the EC_{50} was out of the acceptable range were linked to incorrect media preparation. The mean ± 1 standard deviation of EC_{50}s for assays that were used to determine estrogenic activity of the samples was 2.94 ± 0.71 ng/L.

Results of Blanks and Spikes

Eleven Milli-Q blanks were run through the extraction procedure. Five of these were travel blanks, in which Milli-Q water was sent to the facility in the 1-L amber bottles and returned with the other samples. No estrogenic activity was found in any of the blank samples. Spikes of 2.72 ng/L 17ß-estradiol in Milli-Q or in sample water (matrix spikes) were also assayed with the E-screen (Table 3.1). The Milli-Q water spikes averaged 2.5 ± 0.76 ng/L estrogen activity, and the matrix spikes averaged 2.9 ± 1.1 ng/L activity. The waters from a duplicate bottle of matrix samples without the estrogen spike were all less than the LOQ. Although the standard deviations for these samples are quite high, the results seem reasonable given the variability inherent in bioassays.

Duplicate extractions were completed for six samples (Table 3.2). It was noticed and recorded on the lab slip that the duplicate water for the sample #2 did not resemble the water from its supposed duplicate. Therefore it is not clear what that sample really contained. Excluding duplicate #2, the samples varied between 4 and 31%, and averaged 20% variation.

Table 3.1
E-Screen results from 2.72 ng/L 17β-estradiol-spiked samples.*

Milli-Q water spikes Lab spikes (ng/L)	Milli-Q water spikes Travel spikes (ng/L)	Matrix water spikes (ng/L)
2.78 ± 0.52	1.72 ± 0.30	3.21 ± 0.54
2.42 ± 0.08	4.00 ± 0.22	2.81 ± 0.14
1.74 ± 0.16	2.23 ± 0.14	1.63 ± 0.14
	2.51 ± 0.35	4.63 ± 0.44
		2.29 ± 0.08

*Results represent the mean estrogenic activity ± 1 standard deviation associated with three replicates from a single assay.

Table 3.2
E-Screen results from duplicate samples.*

Duplicate sample	Estrogenic Activity 1 (ng/L)	Estrogenic Activity 2 (ng/L)
1	0.038 ± 0.014	0.046 ± 0.003
2	246 ± 15	1072 ± 40
3	0.120 ± 0.014	0.147 ± 0.003
4	0.087 ± 0.008	0.060 ± 0.003
5	0.133 ± 0.022	0.128 ± 0.005
6	0.060 ± 0.005	0.071 ± 0.003

*Samples were collected in separate 1-L amber bottles and put through the entire extraction and assay procedure separately. Results represent the means ± 1 standard deviation associated with three replicates from a single assay.

Results From Drinking Water Facilities

E-screen assays were conducted on 90 finished water samples from 72 different water treatment facilities. The results from all of the facilities are reported in table C.1. Of the 90 samples tested, the vast majority of the samples (84%) were below the limit of detection (Figure 3.2). With the exception of one facility, all samples were less than 0.27 ng/L of activity. The facility #1016 had an activity of 1.6 ng/L which is of concern as it could possibly be an outlier. Repeating the E-Screen on this finished water would be necessary before having confidence in the result. Human error in extraction or testing, contamination not picked up in the limited number of blanks, or unusual cell activities would need to be ruled out.

In contrast to the finished drinking waters, the majority (61%) of source waters did have measurable estrogenic activity (Figure 3.3). Of the 105 surface or ground water samples tested, 64 were above the detection limit of the assay. Again, the activities are fairly low, with only 10 waters having activities above 0.27 ng/L. Not surprisingly, surface waters exhibited higher levels of estrogenic activity than did ground waters. What was surprising was that 42% the ground waters tested had estrogenic activity. One well that was tested three times (# 1039, well 1) varied from below detection to 0.24 ng/L. The reason for the variation is unknown.

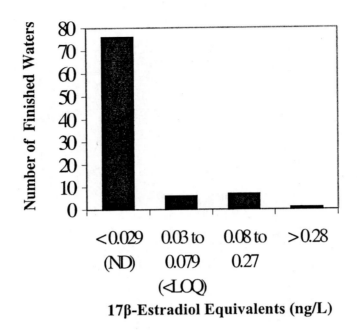

Figure 3.2 Number of finished waters tested with the E-Screen results indicated.

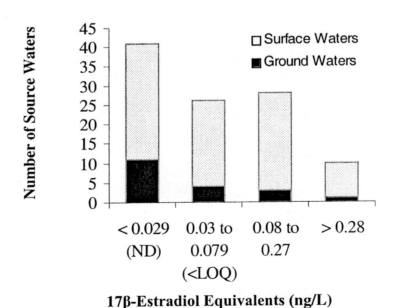

Figure 3.3 Number of source waters tested with the E-Screen results indicated

Drinking water treatment reduced the estrogenic activity in all of the surface waters tested (Table 3.3). In 37 of the samples tested, estrogenic activity was removed to below the limit of detection. According to the US EPA, granular activated carbon (GAC) is a method that can remove endocrine disrupting chemicals from waters (US EPA 2001). GAC removes these organic chemicals by sorption. Although the removal is very effective, total organic carbon needs to be reduced by coagulation/filtration prior to GAC treatment. Additionally, pH and temperature can also alter the removal efficiency. Maintenance of the GAC system is also necessary to ensure removal. Powdered activated carbon (PAC) is generally not as effective as is GAC because less of the water comes into contact with the carbon (USEPA 2001). A few of the facilities where the activity was not as completely removed appeared to employ similar treatment practices to those facilities that were able to remove the activity. For example facility #1017 and #1016 use a common reservoir and the only apparent difference in the treatment train is that #1016 includes activated carbon, which should be able to remove the organic compounds responsible for activity. This facility, however, did not appear to able to remove compounds responsible for the estrogenic activity. It is important to note that each of these sites was only tested one time, and therefore it is not appropriate to draw strong conclusions with such a small sample size. Some facilities use carbon only occasionally, so it is not necessarily the case that it was being used when the sample was collected. Finally, the specific finished water was possibly an outlier, such that the high activity could be due to experimental error.

Table 3.3
E-Screen activity in estradiol equivalents in samples of source and finished waters*

Facility ID	Flow (mgd)	Capacity (mgd)	Population served	Treatment train†	Raw water Eeq (ng/L)‡	Finished water Eeq (ng/L)
1089A	0.19	0.475	1400	CF, S, Car, LS, Cl	2.98 ± 0.04§	ND
1017	75	NA	100000	CF, S, CA, Fl	2.66 ± 0.221	0.191 ± 0.035
1016	95	95	95	CF, S, CA, Car, Fl	1.70 ± 0.095	1.60 ± 0.161
1013	50	100	425000	CF, S, Cl, Fl	1.08 ± 0.033	< LOQ
1001	NA	NA	NA	CF, S, Car, Cl	0.785 ± 0.054	0.212 ± 0.035
1037	1.6	1.4	670	CF, S, Cl, PP	0.343 ± 0.016	0.144 ± 0.003
1012	11.3	55	646000	PP, Aer, Cl	0.294 ± 0.003	ND
1038	12	21	50000	CF, S, Car, Fl, Cl	0.275 ± 0.019	ND
1086	NA	NA	NA	KMn, CF, S, Car, Cl	0.275 ± 0.008	ND
1002	NA	NA	NA	CF, S, PP, Fl, Cl, MF	0.221 ± 0.003	ND
1058	11	20.8	77000	CF, S, Car, CA, Fl, PP, KMn	0.204 ± 0.035	0.120 ±.014§
1026C	95	180	585000	LS, CF, S, Fl, CA	0.188 ± 0.022	< LOQ§
1018	8	32	126000	CF, S, CA	0.183 ± 0.014	0.153 ± 0.014
1064	1.7	2	13200	CF, S, Car, Cl	0.163 ± 0.019	ND
1061	15	18	51000	Car, Fl, KMn, PP, CF, S, Cl	0.161 ± 0.041§	0.087 ± 0.008
1068	7.2	10	33500	Car, Fl, Cl, CF, S	0.133 ± 0.022	ND
1048	70	150	1200000	KMn, CF, S, Fl, CA, O_3, Car	0.125 ± 0.014	ND
1053	12	20	40000	CF, S, Cl, Fl, KMn	0.125 ± 0.011	ND
1024	NA	NA	NA	NA	0.120 ± 0.003	ND
1036	57	200	500000	Car, LS, CF, S, CA, PP	0.120 ± 0.019	ND
1066	4.4	4.8	42325	Car, LS, CF, S, Cl, Fl	0.120 ± 0.014§	ND
1026D	95	180	585000	LS, CF, S, CA, Fl	0.112 ± 0.008	ND
1015A	NA	28.6	NA	RW, Cl	0.101 ± 0.011	ND
1052	7.5	8	30000	CF, S, Car, PP, Fl, Cl	0.101 ± 0.011	ND
1019	24	48	150000	(PP or Car), CF, S, Cl	0.095 ± 0.003	< LOQ
1057	13.5	17	52000	KMn, CF, S, Cl, Car	0.093 ± 0.025§	ND
1014	7.5	12.5	50000	CF, S, Cl, Fl, PP	0.090 ± 0.003	ND
1087	NA	NA	NA	CF, S, O_3, Car, CA	0.084 ± 0.005	ND
1011	NA	NA	70000	LS, Cl	0.082 ± 0.022	ND

(continued)

Table 3.3 (Continued)

Facility ID	Flow (mgd)	Capacity (mgd)	Population served	Treatment train†	Raw water Eeq (ng/L)‡	Finished water Eeq (ng/L)
1054B	17.5	30	130000	CF, S, Car, Fl, CA	0.082 ± 0.011	ND
1026B	95	180	585000	CF, S, LS, CA, Fl	< LOQ	ND
1027	19	32	126000	CF, S, Cl	< LOQ	ND
1035	10.5	23	65000	CF, S, Fl, PP, Cl	< LOQ	ND
1049	30	32	1200000	KMn, CF, S, LS, Fl, CA, PP	< LOQ	< LOQ
1054A	17.5	30	130000	CF, S, Car, Fl, CA	< LOQ	ND§
1065	2.3	3.6	42000	CF, S, Cl, Fl	< LOQ	< LOQ
1067	13.2	16	45000	CF, S, Car, Cl	< LOQ§	ND
1072	NA	230	780000	LS, CF, S, CA, PP, Car, Fl	< LOQ	ND
1073	3.9	6.5	37000	LS, Car, KP, Cl, Fl, CF, S	< LOQ	ND
1074	NA	NA	NA	LS, Ca, Fl	< LOQ	ND§
1076	40	65	176000	CF, S, Car, Fl, Cl	< LOQ	ND
1077	17.4	30	NA	CF, S, Car	< LOQ	ND
1078	20	36	290000	Car, CF, S, Cl	< LOQ§	ND
1079	3	3.1	12500	CF, S, KMn, Cl, Car	< LOQ	ND
1082	8	12	40000	CF, S, KMn, Cl, Car	< LOQ	ND
1083	0.5	2.7	5000	KP, Car, Cl, CF, S, Fl	< LOQ§	ND
1015B	NA	28.6	NA	RW, Cl	< LOQ	ND
1015D	NA	28.6	NA	RW, Cl	< LOQ	ND
1034D	2.5	5.2	20000	CF, S, Cl	< LOQ	ND

*Flow, capacity, population served, and treatment train descriptions are based on information provided by the facilities on the lab slip.
†Abbreviations include NA, not available; O$_3$, ozone contactors; CF, coagulation/flocculation; S, sedimentation followed by sand or anthracite filtration; Car, activated carbon; Fl, flouride addition; Cl, chlorination; CA, chloramination, LS, lime softening; RO, reverse osmosis; PP, polyphosphate; Aer, aeration; KMn, potassium permanganate; RW rainey well collector.
‡Values represent mean ± 1 standard deviation based on 2 or 3 wells in a single assay. ND values are below the limit of detection (0.029 ng/L) and < LOQ values below the limit of quantitation (< 0.079 ng/L) but above the limit of detection.
§Interference in proliferation was indicated by the positive control.

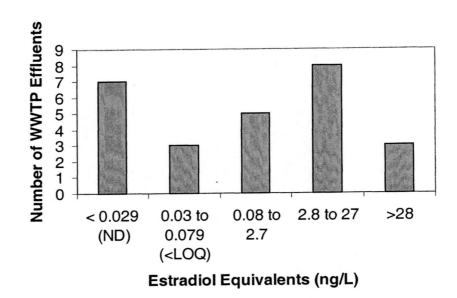

Figure 3.4 Number of wastewater treatment plant effluents tested with the E-Screen results indicated.

Results from Wastewater Treatment Plant Effluents

Twenty-six wastewater treatment plant effluents were tested from 16 facilities. Results are listed in Table (D.1). Estrogenic activity ranged from no activity (in seven of the effluents) to a sample with greater than 500 ng/L of activity (Figure 3.4). Thus, wastewater treatment plant effluents may be responsible for a large amount of estrogenic activity in surface waters. A United States Geological Survey (USGS) in which streams were analyzed for organic wastewater contaminants (including estrogens) documented 17β-estradiol in 10% of the streams surveyed, at levels up to 0.2 μg/L (Rudel et al. 2002). The streams selected for inclusion in the USGS survey were considered likely influenced by WWTP effluents. The detection limit for the analysis used was 0.005 μg/L and therefore the survey likely missed some of the activity that could have been picked up by the E-Screen. In an analysis of groundwater downgradient of an infiltration bed for secondary treated effluent, Kolpin et al. (1998) found bisphenol A and some alkylphenol polyethoxylate metabolites. These compounds are known to have estrogenic activity, and thus sepage from septic systems may explain contamination of groundwater.

The specific processes used in wastewater treatment facilities are very important with respect to the introduction of estrogenic compounds into the environment. Conventional wastewater treatment facilities are not specifically designed to remove these compounds. The findings in the current study are consistent with findings of other researchers in which the efficiencies with which estrogenic compounds are removed vary from nearly complete to ineffective (Ternes et al. 1999; Drewes and Shore, 2001). A few samples were taken at points during treatment and this can provide some insight into effective processes. Secondary and tertiary treatment followed by soil-aquifer treatment and advanced membrane treatment (microfiltration/reverse osmosis) seem to efficiently remove endocrine disrupting activity. Tertiary treatment followed by microfiltration alone appears to be ineffective at removing the estrogen activity (Table 3.4).

Table 3.4
E-Screen activity in estradiol equivalents for various wastewater effluents prior to and after water reclamation processes.

Wastewater followed by water reclamation process	Facility ID	Eeq Prior to treatment (ng/L)*	Eeq After treatment (ng/L)
Secondary Effluent followed by SAT			
-9-18ft below recharge basin receiving secondary effluent	1043	246 ± 15	174 ± 2
-120ft below recharge basin A receiving secondary effluent	1043		ND
-120ft below recharge basin B receiving secondary effluent	1043		< LOQ
Tertiary Effluent followed by SAT			
-70-132ft below ground surface	1044	ND	ND
-100 ft below ground surface /12 months travel time	1042	0.330±0.068	ND
Tertiary Effluent followed by Membrane Treatment			
-after microfiltration	1018	6.24 ± 0.082	7.35 ± 1.53
-after microfiltration plus reverse osmosis	1041	0.518 ± 0.082	ND

*Values represent mean ± 1 standard deviation based on 2 or 3 wells in a single assay. ND values are below the limit of detection (0.029 ng/L) and < LOQ values are below the limit of quantitation (0.079 ng/L) but above the limit of detection.

Evaluation of E-Screen as a Tool

Overall, the E-screen proved to be a good tool to evaluate waters for in vitro estrogenic activity. The responses of the cells were generally consistent, as indicated by the EC50 run chart. Overall, the patterns found seem reasonable; very few of the finished waters exhibited activity relative to the number of surface waters or effluents tested. Thus, even though the number of blanks run was fairly low, the results appear reliable. A few problems were encountered. For several of the samples, the positive control well indicated some type of interference by the sample on the ability of the cells to proliferate. It is not known what caused this. Possibilities include something toxic in the extract, either from the water, residual solvent, or some other contaminant. It is also possible that an estrogen antagonist (compound that binds to the estrogen receptor but prevents estrogen from exerting its effects) was present. Samples in which this occurred may have masked estrogen activity. Further research would be necessary to determine the source of this interference. As indicated earlier, a few samples appeared to be outliers in terms of higher than expected activities relative to other similar waters. This included one finished water. In addition, one sample 1089 was a surface water in which very low activity was expected as it was in a remote area with no impacting WWTP effluents. This sample exhibited a high level of interference in the positive control so the relative low proliferation seen was at the lowest dilution tested. Very few other samples exhibited this level of interference. It is likely that some problem during the extraction procedure occurred to cause this. A repeat sampling at this site the following year did not exhibit any interference in the positive control or estrogenic activity.

FISH STUDIES

Lab Dose Response Study

In the dose response lab exposure to 17β-estradiol, male fathead minnows did induce vitellogenin in a dose dependant manner (Figure 3.5). No vitellogenin was detected in the control or fish exposed to 10 ng/L 17β-estradiol. The GSI was significantly reduced for fish exposed at the 500ng/L estradiol concentration only (data not shown).

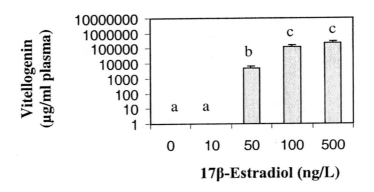

Figure 3.5 Plasma vitellogenin (µg/mL plasma) concentrations for adult male fathead minnows exposed to varying concentrations of estradiol for twenty-one days. Error bars represent standard deviation from three replicate tanks. Responses with different letters are significantly different (α = 0.05) with the SNK means separation test.

Exposure of Fish to Facility Waters

The results of exposures to facility waters are in Table E.1. In the facility water exposures, the responses of the fish to the positive control (100 ng/L dose) varied markedly among the separate studies; the range of vitellogenin induction was between 5900 to 176000 µg/ml plasma. In contrast to the dose response study, variability of fish within individual tanks in the facility water exposures was high. In one positive control tank, the vitellogenin levels varied between 8 and 23,600 µg/ml plasma. Because only one tank was used in this instance, the standard deviation was not calculated for this treatment. In several instances, low level of vitellogenin induction was found in controls or in two instances, in some of the initial fish. As the culture water is recirculated, estrogens excreted by female fish in the culture water may be causing some induction. Although the fish were all within two months of age, differences in responses to the estrogenic compounds may vary based on age, stage of development or other unknown factors.

With two notable exceptions, mortality for the exposures was generally quite low (Table E.1). The fish caged in WWTP effluents for site 1085 and 1088 were lost because cages were lost or destroyed during two separate unusually strong storms. Therefore data is only available at those sites for the lab exposures. Of the facilities tested, two WWTP effluents resulted in significant vitellogenin induction: facilities 1092 and 1088. For site 1092, only the caged fish had significantly elevated vitellogenin levels. The lab-exposed fish had elevated vitellogenin levels, but not significantly so due to large replicate variation. Difference in the lab vs. the field exposures could be due to several reasons including random variability in the responsiveness of the fish used, sorption or degradation of estrogenic compounds during the transport or storage of the sample, or pulses of estrogenic compounds that were missed in the field collections. GSI did not vary in any of the treatments at any of the sites.

The level of estrogen needed to induce vitellogenin in male fathead minnows was in the same range as found by other researcher; 27 ng/L (Kramer et al. 1998) and between 32 and 100 ng/L (Panter et al. 1998). Although vitellogenin induction in males is a valuable tool for determining exposure of organisms to estrogenic compounds, low-level vitellogenin induction does not necessarily translate into adverse reproductive effects. Similar with the findings presented presently, much higher levels of estrogen were needed (320 ng/L) to cause decreases in testicular growth, as indicated by a reduction in GSI (Panter et al. 1998). The concentration needed to reduce egg production in a pair of breeding fathead minnows was 120 ng/L (Kramer et al. 1998, Miles-Richardson et al. 1999), although the variability in this endpoint and problems involved in dosing were substantial, and further work is needed to verify this result.

COMPARISON OF E-SCREEN AND FISH RESPONSES

This study did not prove to be useful in terms of being able to compare the in vitro E-Screen results with the in vivo responses of the fish. Only two samples caused a significant induction in vitellogenin. The samples from facilities 1088 and 1092 had 10.9 and 6.3 ng/L estrogenic activity, respectively in the E-Screen. Based on the lab exposures to estradiol, the fathead minnow would be expected to respond at levels somewhere above 10 ng/L, therefore the vitellogenin response occurred at facility 1092 where it wouldn't have been predicted. Additionally, effluent from facility 1085 was determined by the E-Screen to be the highest of the samples tested with E-Screen and fish simultaneously (12.8 ng/L activity), yet the fish exposed to that effluent did not respond. This was one site where fish cages in field exposures were lost due to severe storms, so it is not known if the field response would have been different. It is possible that the effluents are variable with respect to estrogen activity over a course of a day, or three weeks. If pulses of compounds responsible for activity occur, differences could be due to a one-time sampling of only 1 L, compared to the cumulative effect of the three weeks exposure. Another possible explanation could be a difference in composition of compounds responsible for the activity in effluents. The different compounds could have different sorption properties, differences in bioavailability or requirements for metabolic (in)activation.

CONCLUSIONS

To put this in human terms, at 0.27 ng/L estrogenic activity, someone would have to drink 73420 L of water to get the estrogen potency of one birth control pill (assuming the pill contains 20 µg estrogen and is equivalent to 17β-estradiol). The present analogy notwithstanding, hormones exert their effects at extremely low levels in the body and timing of exposure is critical for proper responses. Many issues remain poorly understood and basic research is still needed because of the complexity of the endocrine system. How compounds with estrogenic activity other than 17 β-estradiol respond in whole organisms will probably not be as simple as looking at their estradiol equivalents in an assay such as the E-Screen. Differences in their solubilities and their ability to bind to plasma binding proteins (Arnold et al. 1996) will likely alter their actual bioavailability. Additionally, their effects could include altering the expression of receptors, or levels of endogenous hormones in the plasma (Geisy et al. 2000). In terms of the exposure of fish to WWTP effluents, the concentration in the highest effluent tested with the E-Screen (579 ng/L) was slightly higher than the concentration that caused 100% mortality (500 ng/L) in one lab exposure and a reduced GSI in the dose response exposure. Thus, although humans likely are exposed to very low levels of estrogenic activity in their drinking water, fish in particular may be experiencing real consequences to exposure of estrogenic compounds in WWTP effluents.

CHAPTER 4
SUMMARY AND CONCLUSIONS

PROJECT DESCRIPTION

Recently, scientific evidence has alerted public health officials, environmental scientists and the drinking water industry that diverse classes of chemicals found in the environment can disrupt the endocrine systems of fish, wildlife, and possibly humans, resulting in measurable, adverse health effects. For example, the decline of the alligator population in Lake Apopka, Florida, was linked to changes in alligator hormone levels following pesticide spills. Hermaphroditic fish have been discovered downstream of sewage-treatment works in the United Kingdom. Male fish collected from a side channel of the Mississippi River have exhibited elevated vitellogenin levels, indicating the presence of estrogenic compounds in this waterway which is used by numerous communities as the source water for their public water supplies. A number of researchers have suggested that declining sperm counts and increasing rates of breast, prostate and testicular cancers in humans may be due to estrogenic compounds found in the environment. Consequently, the USEPA has deemed the study of endocrine disrupters a high priority. It is only a matter of time before regulations concerning the discharge of these compounds are considered. It is imperative that the drinking water community, have as much information as possible about the occurrence and significance of these compounds so as to effectively participate in the regulatory rule making process.

PURPOSE

The purpose of this research was to determine the occurrence, prevalence and significance of estrogenic activity in source waters, finished drinking waters and industrial and municipal wastewater effluents impacting source waters. This goal was accomplished by fulfilling the following objectives.

1. Validate and optimize the E-Screen assay for use with water samples.
2. To document the prevalence of estrogenic activity in waters used as sources for drinking waters, the finished drinking waters, and WWTP effluents.
3. Conduct caged fish studies, in lab exposure studies and the E-Screen assay on the same samples.
4. Evaluate the test results: Assess the reliability of a rapid screen assay compared to in vivo assays. Postulate the possible public health significance of the findings and suggest further research needs.

APPROACH

In the first phase of the research, we modified and validated the E-screen assay to be utilized as an in vitro screening test for the presence of estrogenic compounds in water samples. The E-screen is a cell culture based assay currently used to quantify the occurrence of estrogenic activity of new compounds. The E-screen assay measures in vitro estrogenic activity by the proliferative response of the cells which is mediated by the estrogen receptor. Other in vitro measures were discounted for a variety of reasons during project development. In the second phase of the research, we performed simultaneous in vivo tests, using fathead minnows, and the E-screen to determine the significance of estrogenic compounds present in water. The in vivo fathead minnow test involved exposing caged male minnows in situ, or in aquaria, and testing the minnows for the estrogen-induced protein vitellogenin

SIGNIFICANT RESULTS

Using the E-Screen, we tested 90 finished water samples from 72 facilities. The vast majority of the finished water samples (84%) did not contain estrogenic activity above the level of detection for this assay (0.029 ng/L estradiol equivalents). Of the remaining 14, 13 were below 0.27 ng/L. In contrast to the finished drinking waters, the majority (61%) of source waters did have measurable estrogenic activity. Of the 105 surface or ground water samples tested, 64 were above the detection limit of the assay. Again, the activities were fairly low, with only 10 water having activities above 0.27 ng/L. Not surprisingly, surface waters exhibited higher levels of estrogenic activity than did ground waters. What was surprising was that 42% of the ground waters tested had estrogenic activity. Twenty-six wastewater treatment plant effluents were tested from 16 facilities. In vitro estrogenic activity ranged from no activity (in seven of the effluents) to a sample with greater than 500 ng/L of activity.

Fish exposure studies were completed at six facilities. Vitellogenin levels were significantly increased in male fathead minnows in only two of the waters tested, both in wastewater treatment plant effluents. No significant responses were found in fish exposed to the source or drinking water samples. Comparison between the E-Screen and vitellogenin responses were limited by the low number of positive vitellogenin responses, although based on the E-Screen results, fish would have been expected to respond after exposure to one additional effluent.

CONCLUSIONS

These data indicate that many waters in the U.S. do contain compounds that are estrogenic. This study and others indicate that wastewater treatment plant effluents are one likely source for these compounds. Neither the E-Screen or the induction of vitellogenin can indicate what specific compounds are responsible for this activity. However, natural and synthetic estrogens have been reported to be primarily responsible for estrogenic activity in WWTP effluents. The levels seen in a few of the highly active WWTP effluents in this study may be high enough to affect the health of aquatic organisms. Whether the low levels, such as those seen in a drinking water samples are high enough above human background levels to have an effect is not known. Research is needed to determine if there is a health risk from chronic low levels of estrogenic compounds. Additionally, research is needed to identify specific compounds responsible for the activity and to determine which processes can remove these compounds during treatment at WWTP/and drinking water facilities.

CHAPTER 5
RECOMMENDATIONS FOR THE WATER INDUSTRY

The intent of this study was to develop and evaluate analytical techniques for detecting estrogenically active endocrine disruptors and then apply these techniques to a variety of water sources to start a data base from which the water industry could begin to make decisions related to the occurrence of these compounds. While the adverse health affects to humans from environmental exposure to endocrine disruptors is unknown, it is important for the water industry to proactively engage in the discussion armed with good information. In general terms the study validates the use of both the E-screen assay, and the fathead minnow vitellogenin induction assay for detecting the activity in a variety of water types. This study clearly shows that estrogenically active endocrine disruptors are present in waters important to the drinking water industry. The E-screen tool in particular with an estimated cost of $200 per analysis should prove particularly useful in further characterizations and resolutions of problems related to the occurrence of estrogenically active endocrine disruptors in water.

Previous studies using chemistry analytical techniques have shown that endocrine disruptors occur frequently in wastewaters. This study validates this fact and provides further evidence that these compounds are biologically significant since they did induce vitellogenin production in fathead minnows and produced a cellular response in the E-screen. Watershed managers and utility operators must take this wastewater influence information into consideration when evaluating the impact of these compounds on the operation of their plant.

The other striking piece of information from this study is the surprisingly high percentage of groundwater samples that tested positive. While the levels were generally lower than the surface water supplies, the fact that there were detections at all should serve as a heads-up for operators of groundwater systems. Further work will be required to more fully understand the sources and significance of these detections.

While analysis of the data suggests that removal of estrogenically active endocrine disruptors is occurring during conventional treatment, it is unclear exactly where in the process this removal is taking place. It is also unclear which specific compounds are responsible for the activity. Future work with bench scale jar testing, pilot plant testing and treatment train testing in actual plants will be required. As hormone activity is a biological response, studies using a combination of bioassays and analytical chemical techniques will likely be important in understanding of sources and fate these compounds during treatment.

APPENDIX A
SAMPLE COLLECTION FORM

Biomonitoring

2601 Agriculture Dr. P.O. Box 7996 Madison WI 53707-7996
Tel:608-224-6230 Fax:608-224-6267

Facility

Name:_____ Facility Number:_____
Location: _____ Flow (MGD):_____
Contact: _____ Plant capacity:_____
Phone#: _____ Population served:_____
"Water Type (Source, Finished or Effluent):"_____

Treatment train: attach a schematic or give specific description_____
_____
_____
_____

Chemicals added:_____
Please provide a copy of any available water chemistries on surface and finished waters

Sample

Collector: |_____| Affiliation: _____

Sample Number	Location	Grab	Sample Time MM/DD/YY HH:MM		Temperature (C) In Situ	Sample

For source waters:
Ground or Surface_____ Surface water name:_____
List any pretreatment prior to sampling point: _____
_____
_____

To be completed in the laboratory:
Arrival Temperature (C)_____ pH____D.O_____Cond._____Total Residual
Chlorine (ppm)_____
Chemistries Taken?   Yes   "(Hardness, Ammonium, Alkalinity)"
Leakage?   Yes_____No____ Ice Evident?   Yes___No____

Receiver Signature_____Date/Time_____

Laboratory Number|_____|

extraction date:_____   Escreen date:_____

Disposal: Date/Time_____   Initials _____
Comments:

APPENDIX B
E-SCREEN EC$_{50}$ RUN CHART

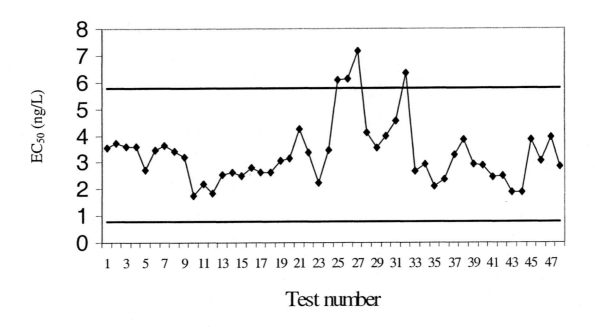

Figure B.1 The run chart of EC_{50} from the E-Screen assays. The diamonds represent EC_{50}s from each assay. The solid lines represent ± 2 standard deviations. Samples were re-run if the EC_{50} did not fall within the 2 standard deviations.

APPENDIX C
E-SCREEN RESULTS FOR SOURCE AND FINISHED WATERS

Table C.1

Facility information and E-screen activity of all source and finished waters tested*

Facility ID	Water type	Flow (mgd)	Capacity (mgd)	Population served	Treatment train†	Estradiol equivalence (ng/L)‡	Coefficient of variation
1001	Surface	NA	NA	NA		0.785 ± 0.054	6.9
1001	Finished	NA	NA	NA	CF, S, Car, Cl	0.212 ± 0.035	16.7
1002	surface	NA	NA	NA		0.221 ± 0.003	1.2
1002	finished	NA	NA	NA	CF, S, PP, Fl, Cl, MF	ND	
1004	surface	66	275	825000		ND	
1004	finished	66	275	825000	O_3, CF, S, Car, Fl, CA	ND	
1005	surface	NA	NA	NA		ND	
1005	finished	NA	NA	NA	O_3	ND	
1005	finished	54	100	825000	O_3, CF, S, Car, Fl, CA	ND	
1006	ground	NA	NA	NA		0.093 ± 0.019	20.6
1006	finished	NA	NA	NA	Cl	ND	

(continued)

Table C.1 (Continued)

Facility ID	Water type	Flow (mgd)	Capacity (mgd)	Population served	Treatment train†	Estradiol equivalence (ng/L)‡	Coefficient of variation
1007	ground					0.346 ± 0.054	15.7
1007	finished	1.474	1.474	9132	LS, RO, Cl, Fl	ND	
1010	surface					ND	
1010	finished	0.4	2	4300	CF, S, Car, Cl	ND	
1011	surface					0.082 ± 0.022	26.7
1011	finished	NA	NA	70000	LS, Cl	ND	
1012	surface					0.294 ± 0.003	0.28
1012	finished	11.3	55	646000	PP, Aer, Cl	ND	
1013	surface					1.08 ± 0.03	3.0
1013	finished	50	100	425000	CF, S, Cl, Fl	<LOQ	
1014	surface					0.090 ± 0.003	3.0
1014	finished	7.5	12.5	50000	CF, S, Cl, Fl, PP	ND	
1016	surface					1.70 ± 0.09	6

(continued)

Table C. 1 (Continued)

Facility ID	Water type	Flow (mgd)	Capacity (mgd)	Population served	Treatment train†	Estradiol equivalence (ng/L)‡	Coefficient of variation
1016	finished	95	NA	100000	CF, S, CA, Car, Fl	1.60 ± 0.16	10
1017	surface	75	NA	100000	CF, S, CA, Fl	2.66 ± 0.22	8.3
1017	finished					0.191 ± 0.035	18.6
1018	surface	8	32	126000	CF, S, CA	0.183 ± 0.014	7
1018	finished					0.153 ± 0.014	9
1019	surface	24	48	150000	(PP or Car), CF, S, Cl	0.095 ± 0.003	3
1019	finished					<LOQ	
1020	surface	NA	120	NA	CF, S, CA, Fl	ND§	
1020	finished					ND§	
1021	surface	NA	80	NA	CF, S, CA, Fl	ND§	
1021	finished					ND§	
1022	surface	23	45	NA	CF, S, CA, Fl	ND	
1022	finished					ND	

(continued)

Table C.1 (Continued)

Facility ID	Water type	Flow (mgd)	Capacity (mgd)	Population served	Treatment train†	Estradiol equivalence (ng/L)‡	Coefficient of variation
1023	surface					ND	
1023	finished	6	14	16000 taps	CF, S, PP, Fl, Cl	ND	
1024	surface					0.120 ± 0.003	2
1024	finished	NA	NA	NA		ND	
1025	surface					ND	
1025	finished	0.25	1.29	2500	S, CA, PP, Fl	ND	
1027	surface					<LOQ	
1027	finished	19	32	126000	CF, S, Cl	ND	
1028	ground					0.120 ± 0.014	11
1028	finished	6	31	NA	None	ND	
1030	ground					ND	
1030	ground					ND	
1030	finished	7	2.2	25000	Cl, PP, Fl	ND	

(continued)

Table C.1 (Continued)

Facility ID	Water type	Flow (mgd)	Capacity (mgd)	Population served	Treatment train†	Estradiol equivalence (ng/L)‡	Coefficient of variation
1031	ground					ND	
1031	ground					ND	
1031	finished	NA	NA	NA	Cl	ND	
1031	finished	NA	NA	NA	Cl	ND	
1033	ground					ND	
1033	ground					ND	
1033	finished	3400 GPM	NA	55000	Cl	ND	
1033	finished	500 GPM	NA	55000	Cl	ND	
1035	surface					<LOQ	
1035	finished	10.5	23	65000	CF, S, Fl, PP, Cl	ND	
1036	surface					0.163 ± 0.025	15.0
1036	surface					0.120 ± 0.019	15.9
1036	finished	57	200	500000	Car, LS, CF, S, CA, PP	ND	

(continued)

Table C.1 (Continued)

Facility ID	Water type	Flow (mgd)	Capacity (mgd)	Population served	Treatment train†	Estradiol equivalence (ng/L)‡	Coefficient of variation
1037	surface						
1037	finished	1150gpm	1.4	670	CF, S, Cl, PP	0.343 ± 0.016	4.7
						0.144 ± 0.003	1.9
1038	surface					0.275 ± 0.019	6.9
1038	finished	12	21	50000	CF, S, Car?, Fl, Cl	ND	
1039C	ground #1					ND	
1039C	ground #2					ND	
1039A	ground #1	NA	2	55000	None	0.242 ± 0.030	12.4
1039B	ground #1			55000		<LOQ	
1039B	ground #2			55000		<LOQ	
1041	surface	28	50	100000		<LOQ§	
1042	finished	185	220	1000000	CF, S, Cl	ND	
1045	surface			61500		ND	
1045	surface	NA	16.8			ND	

(continued)

Table C.1 (Continued)

Facility ID	Water type	Flow (mgd)	Capacity (mgd)	Population served	Treatment train†	Estradiol equivalence (ng/L)‡	Coefficient of variation
1045	finished	7.3	12	61500	Car, CF, S, Cl	ND	
1046	surface					ND	
1046	finished	8	16	20000	Cl	ND	
1047	surface					ND§	
1047	finished	15	15	10000	Dietomeceous Earth Filtration, CA	ND	
1048	surface					0.125 ± 0.014	10.9
1048	finished	70	150	1200000	KMn, CF, S, Fl, CA, O$_3$, Car	ND	
1049	surface					<LOQ	
1049	finished	30	32	1200000	KMn, CF, S, LS, Fl, CA, PP	<LOQ	
1050	surface					ND	
1050	finished	48	86	300000	O$_3$, CF, S, CA, Fl	ND	
1051	surface					ND	
1051	finished	19	48	100000	LS, CF, S, CA, Car, Fl, PP	ND	
1052	surface					0.101 ± 0.011	10.6

(continued)

Table C.1 (Continued)

Facility ID	Water type	Flow (mgd)	Capacity (mgd)	Population served	Treatment train†	Estradiol equivalence (ng/L)‡	Coefficient of variation
1052	finished	7.5	8	30000	CF, S, Car, PP, Fl, Cl	ND	
1053	surface					0.125 ± 0.011	8.7
1053	finished	12	20	40000	CF, S, Cl, Fl, KMn	ND	
1056	surface					ND§	
1056	finished	1	3	3500	CF, S, CA, Fl, PP	<LOQ	
1057	surface	8.9	8.1	NA	Cl, CF, S, Car	0.093§ ± 0.025	26.5
1057	ground	4.8	9	NA	Cl, KMn, CF, S	<LOQ	
1057	finished	13.46	17	52000	Mixed	ND	
1058	surface					0.204 ± 0.035	17.3
1058	finished	11	20.8	77000	CF, S, Car, CA, Fl, PP, KMn	0.120§ ± 0.014	11.4
1059	surface					ND§	
1059	finished	40	56	350000	CF, S, CA, Car, Fl	ND§	
1060	surface					ND§	
1060	finished	10	1	50000	CF, S, CA, Car, Fl	ND§	

(continued)

Table C.1 (Continued)

Facility ID	Water type	Flow (mgd)	Capacity (mgd)	Population served	Treatment train†	Estradiol equivalence (ng/L)‡	Coefficient of variation
1061	surface					0.161§ ± 0.041	25.4
1061	finished	15	18	51000	Car, Fl, KMn, PP, CF, S, Cl	0.087 ± 0.008	9.3
1062	surface					ND	
1062	finished	24	54	250000	Cl, Car, PP, CA, CF, S	ND§	
1063	surface					ND	
1063	finished	10	24	250000	KMn, CF, S, CA, Fl, PP, Car	ND	
1064	surface					0.163 ± 0.019	11.7
1064	finished	1.7	2	13200	CF, S, Car, Cl	ND	
1065	surface					<LOQ	
1065	finished	2.3	3.6	42000	CF, S, Cl, Fl	<LOQ	
1066	surface					0.120§ ± 0.014	11.3
1066	finished	4.4	4.8	42325	Car, LS, CF, S, Cl, Fl	ND	

(continued)

Table C.1 (Continued)

Facility ID	Water type	Flow (mgd)	Capacity (mgd)	Population served	Treatment train†	Estradiol equivalence (ng/L)‡	Coefficient of variation
1067	surface					<LOQ§	
1067	finished	13.2	16	45000	CF, S, Car, Cl	ND	
1068	surface					0.133 ± 0.022	16
1068	finished	7.2	10	33500	Car, Fl, Cl, CF, S	ND	
1070	surface					ND	
1070	finished	11	25	96000	CF, S, CA, Fl	ND	
1072	surface					<LOQ	
1072	finished		230	780000	LD, CF, S, CA, PP, Car, Fl	ND	
1073	surface					<LOQ	
1073	finished	3.9	6.5	37000	LS, Car, KMn, Cl, Fl, CF, S	ND	
1074	surface					<LOQ	
1074	finished	NA	NA	NA	LS, Ca, Fl	ND§	
1075	surface					ND	
1075	finished	7.48	12	56903	CF, S, KMn, PP, Cl	ND	

(continued)

Table C.1 (Continued)

Facility ID	Water type	Flow (mgd)	Capacity (mgd)	Population served	Treatment train†	Estradiol equivalence (ng/L)‡	Coefficient of variation
1076	surface					<LOQ	
1076	finished	40	65	176000	CF, S, Car, Fl, Cl	ND	
1077	surface					<LOQ	
1077	finished	17.4	30	NA	CF, S, Car	ND	
1078	surface					<LOQ§	
1078	finished	20	36	290000	Car, CF, S, Cl	ND	
1079	surface					<LOQ	
1079	finished	3	3.1	12500	CF, S, KMn, Cl, Car	ND	
1080	surface					ND§	
1080	finished	20	30	39000	KMn, CF, S, CA, Car, Fl	ND	
1082	surface					<LOQ	
1082	finished	8	12	40000	CF, S, KMn, Cl, Car	ND	

(continued)

43

Table C.1 (Continued)

Facility ID	Water type	Flow (mgd)	Capacity (mgd)	Population served	Treatment train†	Estradiol equivalence (ng/L)‡	Coefficient of variation
1083	surface	0.5	2.7	5000	KMn, Car, Cl, CF, S, Fl	<LOQ§	
1083	finished					ND	
1084	surface					ND	
1084	surface					0.185 ± 0.041	22.1
1084	surface					0.084 ± 0.003	4.5
1086	surface	NA	NA	NA	KMn, CF, S, Car, Cl	0.275 ± 0.008	3.0
1086	finished					ND	
1087	surface					0.084 ± 0.005	7.1
1087	finished				CF, S, O$_3$, Car, CA	ND	
1015A	surface					0.120 ± 0.008	6.8
1015A	surface					0.101 ± 0.011	10.8
1015A	finished	NA	28.6	NA	RW, Cl	ND	
1015B	surface					<LOQ	18.5
1015B	surface					<LOQ	35.7

(continued)

Table C.1 (Continued)

Facility ID	Water type	Flow (mgd)	Capacity (mgd)	Population served	Treatment train†	Estradiol equivalence (ng/L)‡	Coefficient of variation
1015B	finished	NA	28.6	NA	RW, Cl	ND	
1015C	surface						
1015C	surface						
1015C	finished	NA	28.6	NA	RW, Cl	ND	
1015D	surface					<LOQ	
1015D	surface					<LOQ	
1015D	finished	NA	28.6	NA	RW, Cl	ND	
1026A	surface					ND	
1026A	finished	95	180	585000	CF, S, LS, Fl, CA	ND	
1026A	ground					ND	
1026A	finished	30	60	195000	RW, CF, S, LS, Fl, CA	ND	
1026B	surface					<LOQ	
1026B	finished	95	180	585000	CF, S, LS, Fl, CA	ND	
1026B	ground					ND	
1026B	finished	30	60	195000	RW, CF, S, LS, Fl, CA	ND	

(continued)

Table C.1 (Continued)

Facility ID	Water type	Flow (mgd)	Capacity (mgd)	Population served	Treatment train†	Estradiol equivalence (ng/L)‡	Coefficient of variation
1026C	surface					0.188 ± 0.022	11.6
1026C	finished	95	180	585000	CF, S, LS, Fl, CA	<LOQ§	
1026C	ground					<LOQ§	31.8
1026C	finished	30	60	195000	RW, CF, S, LS, Fl, CA	0.120§ ± 0.038	
1026D	surface					0.112 ± 0.008	7.3
1026D	finished	95	180	585000	CF, S, LS, Fl, CA	ND	
1026D	ground					ND	
1026D	finished	30	60	195000	RW, CF, S, LS, Fl, CA	ND	
1034A	surface					ND§	
1034A	finished	3.8	5.2	20000	CF, S, Cl	ND§	
1034B	surface					ND	
1034B	finished	4.27	5.2	20000	CF, S, Cl	ND	
1034C	surface					ND	
1034C	finished	2.5	5.2	20000	CF, S, Cl	ND	

(continued)

Table C.1 (Continued)

Facility ID	Water type	Flow (mgd)	Capacity (mgd)	Population served	Treatment train†	Estradiol equivalence (ng/L)‡	Coefficient of variation
1034D	surface					<LOQ	
1034D	finished	2.5	5.2	20000	CF, S, Cl	ND	
1054A	surface					<LOQ	
1054A	finished	17.5	30	130000	CF, S, Car, Fl, CA	ND§	
1054B	surface					0.082 ± 0.011	13
1054B	finished	17.5	30	130000	CF, S, Car, Fl, CA	ND	
1089A	surface					2.98§ ± 0.03	
1089A	finished	0.19	0.475	1400	CF, S, Car, LS, Cl	ND	
1089B	surface					ND	
1089B	finished	0.19	0.475	1400	CF, S, Car, LS, Cl	ND	

*Flow, capacity, population served, and treatment train descriptions are based on information provided by the facilities on the lab slip.
†Abbreviations include NA, not available; O_3, ozone contactors; Cl, chlorination; CA, chloramination; LS, lime softening; RO, reverse osmosis; CF, coagulation/flocculation; S, sedimentation followed by sand or anthracite filtration; Car, carbon; Fl, flouride addition; PP, polyphosphate; Aer, aeration; KMn, potassium permanganate; RW, rainey well collector.
‡Values represent mean ± 1 standard deviation based on 2 or 3 wells in a single assay. ND values are below the limit of detection (0.029 ng/L) and <LOQ values below the limit of quantitation (0.079 ng/L) but above the limit of detection.
§Interference in proliferation was indicated by the positive control.

APPENDIX D
E-SCREEN RESULTS FOR WWTP EFFLUENTS

Table D.1
Facility information and E-screen activity of all wastewater treatment plant effluents tested*

Facility ID	Flow (mgd)	Capacity (mgd)	Population served	Treatment train†	Estradiol equivalents (ng/L)‡	Coefficient of variation (%)
1008	NA	1.36	NA	Pri, Sec, Ter, Cl, Dec	5.56 ± 0.93	16.4
1009	NA	NA	NA	NA	2.16 ± 0.17	8.1
1018	12	43	NA	Pri, Sec, Ter	6.25 ± 0.09	1.4
1018				Pri, Sec, Ter, MF	7.38 ± 1.53	20.7
1035	NA	3.25	NA	Pri, Sec, UV	0.586 ± 0.044	7.4
1040	12	25	50000	Pri, Sec, Cl, Dec	17.4 ± 2.3	13.1
1041	16	12	100000	Pri, Sec, Ter, CA	0.528 ± 0.076	14.4
1041	10	10	100000	Pri, Sec, Ter, CA, RO, MF	ND§	
1042	12	16	100000	Pri, Sec, Ter, UV	0.286 ± 0.008	2.9
1042				Pri, Sec, Ter, UV, SAT	ND	

(continued)

Table D.1 (Continued)

Facility ID	Flow (mgd)	Capacity (mgd)	Population served	Treatment train†	Estradiol equivalents (ng/L)‡	Coefficient of variation (%)
1043	25	32	NA	Pri, Sec, Cl, Dec	246 ± 15	6.1
1043				Pri, Sec, 9-18ft SAT	174 ± 2	1.2
1043				Pri, Sec, 120 ft SAT	<LOQ	
1043				Pri, Sec, 120 ft SAT	ND	
1044	58	62.5	574000	Pri, Sec, Tert	ND	
1044				Pri, Sec, Ter, SAT	ND	
1047	NA	16	10000	Pri, Sec, Ter	<LOQ	
1085	NA	NA	NA	Pri, Sec, UV	12.8 ± 0.3	2.4
1088	NA	NA	NA		11.1 ± 1.9	17.5
1090	1	NA	NA	Pri, Sec, OD, Cl	579 ± 14	2.5

(continued)

Table D.1 (Continued)

Facility ID	Flow (mgd)	Capacity (mgd)	Population served	Treatment train†	Estradiol equivalents (ng/L)‡	Coefficient of variation (%)
1092	NA	NA	NA	Pri, Sec	6.21 ± 0.37	5.8
1093	NA	NA	NA	Pri, Sec, Cl, Dec	4.14 ± 0.18	4.3
1034A	2.11	4.5	20000	Pri, Sec, Cl, Dec	0.199 ± 0.035	18
1034B	2.05	4.5	20000	Pri, Sec, Cl, Dec	<LOQ	
1034C	1.61	4.5	20000	Pri, Sec, Cl, Dec	ND§	
1034D	1.61	4.5	20000	Pri, Sec, Cl, Dec	ND§	

*Flow, capacity, population served, and treatment train descriptions are based on information provided by the facilities on the lab slip.
†Abbreviations include NA, not available; Pri, primary settling; Sec, secondary biological treatment; Ter, tertiary treatment; Fil, filtration; Cl, chlorination; Dec, dechlorination; MF, microfiltration; UV ultra violet disinfection; OD, oxidation ditch; SAT, soil aquifer treatment.
‡Values represent mean ± 1 standard deviation based on 2 or 3 wells in a single assay. ND values are below the limit of detection (0.029 ng/L) and <LOQ values below the limit of quantitation (0.079 ng/L) but above the limit of detection.
§Interference in proliferation was indicated by the positive control.

APPENDIX E
MALE FATHEAD MINNOW EXPOSURES

Table E.1
Responses of male fathead minnows to different waters

STUDY	Treatment (facility number)	N*	Lab or field	Mortality (%)	GSI† (%)	Vitellogenin (µg/ml plasma)
1	Initial fish	4	Lab	0	1.51 ± 0.15	0.00 ± 0.00 a
	Lab control	3	Lab	0	2.06 ± 0.50	0.60 ± 0.60 a
	Lab control in cubitainer	3	Lab	0	1.46 ± 0.06	1.00 ± 1.00 a
	Lab control + sodiumthiosulfate	2	Lab	0	1.29 ± 0.01	0.00 ± 0.00 a
	Raw drinking water (1001)	3	Field	13	1.63 ± 0.20	0.00 ± 0.00 a
	Raw drinking water (1001)	3	Lab	6	1.65 ± 0.08	0.00 ± 0.00 a
	Finished drinking water (1001)	3	Lab	6	1.64 ± 0.11	0.00 ± 0.00 a
	WWTP effluent (1092)	3	Field	7	1.76 ± 0.15	560 ± 193 b
	WWTP effluent (1092)	3	Lab	0	1.78 ± 0.06	171 ± 109 a
	Lab control + 500 ng/L estradiol	1	Lab	100	--	--
1	P-value				0.550	< 0.001
2	Initial fish	4	Lab	0	1.45 ± 0.11	0.00 ± 0.00 a
	Lab control	3	Lab	0	1.56 ± 0.07	20 ± 6 a
	Lab control in cubitainer	3	Lab	6	1.62 ± 0.13	14.1 ± 1.1 a
	Lab control + sodiumthiosulfate	2	Lab	0	1.23 ± 0.33	0.00 ± 0.00 a
2	Raw drinking water (1002)	3	Field	3	1.84 ± 0.34	0.00 ± 0.00 a
	Raw drinking water (1002)	3	Lab	0	1.42 ± 0.09	0.00 ± 0.00 a
	Finished drinking water (1002)	3	Lab	22	1.60 ± 0.14	3.3 ± 3.3 a

(continued)

Table E.1. (Continued)

STUDY	Treatment (facility number)	N*	Lab or field	Mortality (%)	GSI† (%)	Vitellogenin (μg/ml plasma)
	WWTP effluent (1093)	3	Field	77	1.65 ± 0.48	0.00 ± 0.00 a
	WWTP effluent (1093)	3	Lab	0	1.84 ± 0.21	6.4 ± 5.2 a
	Lab control + 100 ng/L estradiol	1	Lab	0	0.92	176000 b
2	P-value				0.694	< 0.001
3	Initial fish	8	Lab	0	1.39 ± 0.10	1.2 ± 1.2 a
	Lab control	3	Lab	0	1.59 ± 0.12	1.3 ± 0.7 a
	Lab control in cubitainer	3	Lab	0	1.58 ± 0.11	2.4 ± 2.1 a
	Lab control + sodiumthiosulfate	2	Lab	0	1.61 ± 0.09	0.00 ± 0.00 a
	Raw drinking water (1089)	3	Field	0	1.88 ± 0.06	0.00 ± 0.00 a
	Raw drinking water (1089)	3	Lab	0	1.83 ± 0.09	0.00 ± 0.00 a
	Finished drinking water (1089)	3	Lab	0	1.77 ± 0.22	3.6 ± 3.0 a
	Lab control + 100 ng/L estradiol	1	Lab	17	1.3	89100 b
3	P-value				0.283	< 0.001
4	Initial fish	9	Lab	0	1.10 ± 0.20	0.00 ± 0.00 a
	Lab control + sodiumthiosulfate	2	Lab	0	1.23 ± 0.32	0.00 ± 0.00 a
	Raw drinking water (1086)	3	Field	100	NA	NA
	Raw drinking water (1086)	3	Lab	0	1.19 ± 0.13	0.00 ± 0.00 a

(continued)

Table E.1. (Continued)

STUDY	Treatment (facility number)	N*	Lab or field	Mortality (%)	GSI† (%)	Vitellogenin (μg/ml plasma)
	Finished drinking water (1086)	3	Lab	6	1.31 ± 0.17	0.00 ± 0.00 a
	WWTP effluent (1085)	3	Field	100	--	--
	WWTP effluent (1085)	3	Lab	6	1.43 ± 0.20	5.1 ± 5.1 a
	Lab control + 100 ng/L estradiol	2	Lab	0	1.33 ± 0.15	9100 ± 2300 b
4	P-value				0.875	< 0.001
5	Initial fish	9	Lab	0	1.87 ± 0.21	0.00 ± 0.00 a
	Lab control in cubitainer	3	Lab	0	1.68 ± 0.11	2 ± 2 a
	Lab control + sodiumthiosulfate	2	Lab	0	1.73 ± 0.00	0.00 ± 0.00 a
	Raw drinking water (1087)	3	Field	23	1.96 ± 0.26	0.00 ± 0.00 a
	Raw drinking water (1087)	3	Lab	0	1.84 ± 0.08	0.00 ± 0.00 a
	Finished drinking water (1087)	3	Lab	0	1.77 ± 0.08	0.00 ± 0.00 a
	WWTP effluent (1088)	3	Lab	0	1.37 ± 0.03	12000 ± 3000 b
	WWTP effluent (1088)	3	Field	100	--	--
	Lab control + 100 ng/L estradiol	2	Lab	0	1.82 ± 0.2	24000 ± 4000 c
5	P-value				0.183	< 0.001
6	Initial fish	8	Lab	0	1.75 ± 0.21	1.50 ± 1.50 a
	Lab control in cubitainer	3	Lab	6	1.81 ± 0.8	0.00 ± 0.00 a
	Lab control + sodiumthiosulfate	2	Lab	6	1.55 ± 0.16	0.00 ± 0.00 a
	Raw drinking water (1054)	3	Field	7	1.69 ± 0.22	8 ± 8 a

(continued)

Table E.1. (Continued)

STUDY	Treatment (facility number)	N*	Lab or field	Mortality (%)	GSI† (%)	Vitellogenin (µg/ml plasma)
	Raw drinking water (1054)	3	Lab	0	1.65 ± 0.09	0.00 ± 0.00 a
	Finished drinking water (1054)	3	Lab	0	2.25 ± 0.50	0.00 ± 0.00 a
	Lab control + 100 ng/L estradiol	2	Lab	0	1.45 ± 0.00	5900 ± 900 b
6	P-value				0.526	< 0.001

*N represents the number of replicates, which is the number of tanks or cages for lab and field exposures, respectively. An exception is for the initial fish, which indicates the actual number of fish.
†Responses for gonadosomatic index (GSI) and vitellogenin induction are based on the mean response from replicate tanks or cages. P-values represent the results of an analysis of variance within a site. In studies with significant treatment effects, responses with different letters are significantly different ($\alpha = 0.05$) with the SNK means separation test.

REFERENCES

Arnold, S.F., M.K. Robinson, A.C. Notides, L.J Guillette, and J.A. McLachlan. 1996. A Yeast Estrogen Screen for Examining the Relative Exposure of Cells to Natural and Xenoestrogens. Environmental Health Perspectives, 104(5):544-548.

APHA, AWWA, and WEF (American Public Health Association, American Water Works Association, and Water Environment Federation). 1992. *Standard Methods for the Examination of Water and Wastewater.* 18th ed. Washington, D.C.: APHA

Colborn, R., D. Dumanoski, and J.P. Myers. 1996. *Our Stolen Future.* New York: Plume.

Crisp, T.M., E.D. Clegg, R.L. Cooper, W.P. Wood, D.G. Anderson, K.P. Baetcke, J.L. Hoffmann, M.S. Morrow, D.J. Rodier, J.E. Schaeffer, L.W. Touart, M.G. Zeeman, and Y.M. Patel. 1998. Environmental Endocrine Disruption: an Effects Assessment and Analysis. *Environmental Health Perspectives,* 106(suppl 1):11-56.

Desbrow, C., E.J. Routledge, G.C. Brighty, J.P. Sumpter, and M. Waldock, M. 1998. Identification of Estrogenic Chemicals in STW Effluent. 1. Chemical Fractionation and in Vitro biological screening. *Environmental Science and Technology,* 32(11):1549-1558.

Drewes, J.E. and Shore, L.S. (2001). Concerns about pharmaceuticals in water reuse, groundwater recharge, and animal waste. In *Pharmaceuticals and Personal Care Products in the Environment: Scientific and Regulatory Issues* Edited by C.G. Daughton, C.G. and T.Jones-Lepp. Symposium Series 791. Washington, D.C.: American Chemical Society.

EDSTAC (Endocrine Disruptor Screening and Testing Advisory Committee) 1998. Final report. (http://www.epa.gov/scipoly/oscpendo/history/finalrpt.htm). U.S. EPA: Washington D.C.

Folmar, L.D., N.D. Denslow, K. Kroll, E.F. Orlando, J. Enblom, J. Marcino, C. Metcalfe, and L.J. Guillette. 2001 Altered Serum Sex Steroids and Vitellogenin Induction in Walleye *(Stizostedion vitreum)* Collected Near a Metropolitan Sewage Treatment Plant. *Archives of Environmental Contamination and Toxicology,* 40(3):392-398.

Folmar LC, N.D. Denslow, V. Rao, M. Chow, D.A. Crain, J. Enblom, J. Marcino, and L.J. Jr., Guillette. 1996. Vitellogenin induction and reduced serum testosterone concentrations in feral male carp (Cyprinus carpio) captured near a major metropolitan sewage treatment plant. *Environmental Health Perspectives,* 104:1096-1101.

Giesy, J.P., S.L. Pierens, E.M. Snyder, S. Miles-Richardson, V.J. Kramer, S.A. Snyder, K.M. Nichols, and D.A. Villeneuve. 2000. Effects of 4-nonylphenol on fecundity and biomarkers of estrogenicity in fathead minnows (Pimephalas promelas*). Environmental Toxicology and Chemistry,* 19(5):1368-1377.

Guillette, L.J., Jr., T.S. Gross, G.R. Masson, M.M. .Matter, H.F. Percival, and A.R. Woodward 1994. Developmental abnormalities of the gonad and abnormal sex hormone concentrations in juvenile alligators from contaminated and control lakes in Florida. *Environmental Health Perspectives,* 102:680-688.

Guillette, L.J., Jr., D.B. Pickford, D.A. Crain, A.A. Rooney, and H.F. Percival. 1996. Reduction in penis size and plasma testosterone concentrations in juvenile alligators living in a contaminated environment. *General and Comparative Endocrinology,* 101:32-42.

Handelsman, D.J. 2001. Estrogens and falling sperm counts. *Reproduction, Fertility, and Development,* 13(4):317-324.

Harries, J.E., D.A. S. Sheahan, P. Jobling, P. Matthiessen, E.J. Neall, R. Routledge, J.P. Rycroft, J.P. Sumpter, and T. Tylor. 1996. A survey of estrogenic activity in United Kingdom inland waters. *Environmental Toxicology and Chemisty,* 15(11):1993-2002.

Harries, J.E., D.A. S. Sheahan, P. Jobling, P. Matthiessen, E.J. Neall, R. Routledge, J.P. Rycroft, J.P. Sumpter, T. Tylor, and N. Zaman. 1997. Estrogenic activity in five United Kingdom rivers detected by measurement of vitellogenesis in caged male trout. *Environmental Toxicology and Chemisty,* 16:534-542.

Jensen, T.K., J. Toppari, N. Keiding, and N.E. Skakkebaek. 1995. Do Environmental Estrogens Contribute to the Decline in Male Reproductive Health? *Clinical Chemistry,* 41(12B Part 2 Suppl S):1896-1901.

Jobling, S., D. Sheahan, J.A. Osborne, P. Matthiessen, and J.P. Sumpter. 1996. Inhibition of Testicular Growth in Rainbow Trout (*Oncorhynchus mykiss*) Exposed to Estrogenic Alkylphenolic Chemicals. *Environmental Toxicology and Chemistry,* 15:194-202.

Jobling, S., and J.P. Sumpter. 1993. Detergent Components in Sewage Effluent are Weakly Oestrogenic to Fish: An In Vitro Study Using Rainbow Trout (Oncorhynchus mykiss) hepatocytes. *Aquatic Toxicology,* 27:361-372.

Jobling, S., M. Nolan, C.R. Tyler, G. Brighty, and J.P. Sumpter. 1998. Widespread Sexual Disruption in Wild Fish. *Environmental Science and Technology,* 32:2498-2506.

Jobling, S., T. Reynolds, R. White, M.G. Parker, and J.P. Sumpter. 1995. A Variety of Environmentally Persistent Chemicals, Including Some Phthalate Plasticizers, are Weakly Estrogenic. Environmental Health Perspectives, 103(6):582-587.

Jones, P.D., W.M. De Coen, L. Tremblay, and J.P. Giesy. 2000. Vitellogenin as a Biomarker for Environmental Estrogens. *Water Science and Technology,* 42(7-8):1-14.

Knudsen, F.R., A.E. Schou, M.L. Wiborg, E. Mona, K-E, Tollefsen, J. Stenersen, and J.P. Sumpter. 1997. Increase of Plasma Vitellogenin Concentration in Rainbow Trout (*Oncorhynchus mykiss*) Exposed to Effluents From oil Refinery Treatment Works and Municipal Sewage. *Bulletin of Environmental Contamination and Toxicology,* 59:802-806.

Kolpin, D.W., E.T. Furlong, M.T. Meyer, E.M. Thurman, S.D. Zaugg, L.B. Barber, and H.T. Buxton. 2002. Pharmaceuticals, Hormones, and Other Organic Wastewater Contaminants in U.S. Streams, 1999-2000: A National Reconnaissance. *Environmental Science and Technology,* 36(6): 1202-1211.

Kramer, V.J., S. Miles-Richardson, S.L. Pierens, and J.P. Giesy. 1998. Reproductive Impairment and Induction of Alkaline-Labile Phosphate, a Biomarker of Estrogen Exposure, in Fathead Minnows. *Aquatic Toxicology,* 40:335-360.

Lye, C.M., C.L.J. Frid, and M.E. Gill. 1998. Seasonal Reproductive Health of Flounder *Platichthys flesus* Exposed to Sewage Effluent. *Marine Ecology Progress Series,* 170:249-260.

Matsui, S., H. Takigami, T. Matsuda, N. Taniguchi, J. Adachi, H. Kawami, and Y. Shimizu. 2000. Estrogen and Estrogen Mimics Contamination in Water and the Role of Sewage Treatment. *Water Science and Technology,* 42(12):173-179.

Mellanen, P., T. Petänen, J. Lehtimāke, S. Mäkelä, G. Bylund, B. Holmbom, E. Mannila, A. Oikari, and R. Santti. 1996. Wood-Derived Estrogens: Studies *in Vitro* With Breast Cancer Cell Lines and *in Vivo* in Trout. *Toxicology and Applied Pharmacology,* 136:381-388.

Miles-Richardson, S.R., V.J. Kramer, S.D. Fitzgerald, J.A. Render, B. Yamini, S.J. Barbee, and J.P. Geisy. 1999. Effects of Waterborne Exposure of 17 β-estradiol on Secondary sex Characteristics and Gonads of Fathead Minnows (*Pimephalas promelas*). *Aquatic Toxicology*, 47:129-145.

Panter, G.H., R.S. Thompson, and J.P. Sumpter. 1998. Adverse Reproductive Effects in Male Fathead Minnows (*Pimephales promelas*) Exposed to Environmentally Relevant Concentrations of the Natural Oestrogens, Oestradiol and Oestrone. *Aquatic Toxicology*, 42:243-253.

Payne, J., C. Jones, S. Lakhani, and A. Kortenkamp. 2000. Improving the reproducibility of the MCF-7 cell proliferation assay for the detection of xenoestrogens. *The Science of the Total Environment*, 248:51-62.

Roefer, P., S. Snyder, R.E. Zegers, D.J. Rexing, and J.L. Fronk. 2000. Endocrine-disrupting Chemicals in a Source Water. *Journal Awwa*, 92(8):52-58.

Routledge, E., J. and J.P. Sumpter. 1996. Estrogenic Activity and Some of their Degradation Products Assessed Using a Recombinant Yeast Screen. *Environmental Toxicology and Chemistry* 15(3):241-248.

Rudel, R. A., S. J. Melly, P.W. Geno, G. Sun, and J.G. Brody, J. G. 1998. Identification of Alkylphenols and other Estrogenic Phenolic Compounds in Wastewater, Septage, and Groundwater on Cape Cod, Massachusetts. *Environmental Science & Technology* 32: 861-869.

Safe, S.H. 2000. Endocrine Disruptors and Human Health – Is there a Problem? An Update. *Environmental Health Perspectives*, 108(6):487-493.

Sharpe R.M., and N.E. Skakkebaek. 1993 Are oestrogens involved in falling sperm counts and disorders of the mail reproductive tract? *Lancet,* 341:1392-1395.

Snyder, S.A., D.L. Villeneuve, E.M. Snyder, and J.P. Giesy. 2001. Identification and Quantification of Estrogen Receptor Agonists in Wastewater Effluents. *Environmental Science and Technology*, 35(18):3620-3625.

Snyder, S. A., T.L. Keith, D.A. Verbrugge, E.M. Snyder, T.S. Gross, K. Kannan K., and J.P. Giesy. 1999. Analytical Methods for Detection of Selected Estrogenic Compounds in Aqueous Mixtures. *Environmental Science & Technology*, 33: 2814-2820.

Soimasuo, M.R., A..E. Karels, H. Leppänen, R. Santti, and A.O.J. Oikari. 1998. Biomarker responses in whitefish (*Coregonus lavaretus* L. s.l.) experimentally exposed in a large lake receiving effluents from pulp and paper industry. *Archives of Environmental Contamination and Toxicology*, 34:69-80.

Soto, A., C. Sonnenschein, K. Chung, M. Fernandez, N. Olea, and F. Olea Serrano. 1995. The E-SCREEN Assay as a Tool to Identify Estrogens: an Update on Estrogenic Environmental Pollutants. *Environmental Health Perspectives,* 103 (Suppl. 7):113-122.

Sumpter, J.P., and S. Jobling. 1995. Vitellogenesis as a biomarker for estrogenic contamination of the aquatic environment. *Environmental Health Perspectives,* 103 (Suppl. 7):173-178.

Ternes, T.A., Stumpf, M., Mueller, J., Haberer, K., Wilken, R.-D., and Servos, M. 1999. Behavior and occurrence of estrogens in municipal sewage treatment plants – I. Investigations in Germany, Canada and Brazil. *The Science of the Total Environment* 225: 81-90.

Tyler, C.R., S. Jobling, and J.P. Sumpter. 1998. Endocrine Disruption in Wildlife: a Critical Review of the Evidence. *Critical Reviews in Toxicology,* 28:319-361.

U.S. Environmental Protection Agency. 1997. Update to the Office of Research and Development's Strategic Plan. Washington D.C.: EPA/600/R-97/015.

U.S. Environmental Protection Agency. 1987. Guidelines for the culture of fathead minnows Pimephales promelas for use in toxicity tests. Washington D.C.:E PA/600/3-87/001.

U.S. Environmental Protection Agency. 2001. Removal of Endocrine Disruptor Chemicals Using Drinking Water Treatment Processes. Washington D.C.:E PA/625/R-00/015.

Zacharewski T. In Vitro Bioassays for Assessing Estrogenic Substances. 1997. *Environmental Science and Technology*, 31:613-623.

ABBREVIATIONS

Awwa	American Water Works Association
AwwaRF	Awwa Research Foundation
°C	degrees Celsius
CD	charcoal dextran stripped media
CD FBS	charcoal dextran stripped fetal bovine serum
cm^2	square centimeter
CO_2	carbon dioxide
DCM	dichloromethane
dechlor	dechlorinated water
DO	dissolved oxygen
EC_{50}	The concentration that causes 50% of the maximum effect
Eeq	Estradiol equivalents
ELISA	Enzyme linked immunosorbent assay
°F	degrees Fahrenheit
FBS	fetal bovine serum
g	grams
GAC	granular activated carbon
h	hour
L	liter
LOD	Limit of detection
LOQ	limit of quantitation
µg	microgram
µm	micrometer
M	molar (moles/L)
mg	milligram
mL	milliliter
ND	No detect, below limit of detection
ng	nanogram
nm	nanometer
OD	optical density
pM	picomolar

SRB	sulphorhodamine B
TCA	tricholor acetic acid
WTP	water treatment plant
WWTP	wastewater treatment plant